How to Cope with Climate Change

Other World Scientific Books on Climate Change

Waking Up to Climate Change:
Five Dimensions of the Crisis and What We Can Do About It
by George Ropes
ISBN: 978-981-12-4623-4
ISBN: 978-981-12-4754-5 (pbk)

Decarbonizing Asia: Innovation, Investment and Opportunities
edited by Tony Á Verb and Roman Y Shemakov
contributions by Alexandra Tracy, Sandro Desideri, Eric Chong,
Bill Kentrup, James Kruger, Moon K Kim, Alberto Balbo,
Christine Loh, Archawat Chareonsilp, Suede Kam, Rachel Fleishman, Amarit Charoen-
phan and Davide A Nicolini
ISBN: 978-981-12-6386-6
ISBN: 978-981-12-6466-5 (pbk)

Science in Society: Climate Change and Climate Policies
by Nico Stehr and Hans von Storch
ISBN: 978-1-80061-351-5

Peace with Nature: 50 Inspiring Essays on Nature and the Environment
edited by Tommy Koh, Lin Heng Lye and Shawn Lum
ISBN: 978-981-12-8201-0
ISBN: 978-981-12-8272-0 (pbk)

The Earth is Not for Sale: A Path Out of Fossil Capitalism to the
Other World That is Still Possible
by Peter Schwartzman and David Schwartzman
ISBN: 978-981-323-424-6
ISBN: 978-981-327-664-2 (pbk)

The Energy Conundrum: Climate Change, Global Prosperity, and the Tough
Decisions We Have to Make
by Neil A C Hirst
ISBN: 978-1-78634-460-1
ISBN: 978-1-78634-667-4 (pbk)

Co-Existing with the Earth: Tzu Chi's Three Decades of Recycling
by William Kazer and Rey-Sheng Her
ISBN: 978-981-12-3223-7
ISBN: 978-981-12-3157-5 (pbk)

The Goldilocks Policy: The Basis for a Grand Energy Bargain
by John R Fanchi
ISBN: 978-981-327-639-0
ISBN: 978-981-327-744-1 (pbk)

Saving Ourselves: Interviews with World Leaders on the Sustainable Transition
edited by Yacine Belhaj-Bouabdallah
ISBN: 978-981-322-074-4
ISBN: 978-981-322-075-1 (pbk)

How to Cope with Climate Change

Michael Krause

W World Scientific

NEW JERSEY · LONDON · SINGAPORE · BEIJING · SHANGHAI · HONG KONG · TAIPEI · CHENNAI · TOKYO

Published by

World Scientific Publishing Co. Pte. Ltd.

5 Toh Tuck Link, Singapore 596224

USA office: 27 Warren Street, Suite 401-402, Hackensack, NJ 07601

UK office: 57 Shelton Street, Covent Garden, London WC2H 9HE

Library of Congress Control Number: 2024000600

British Library Cataloguing-in-Publication Data
A catalogue record for this book is available from the British Library.

HOW TO COPE WITH CLIMATE CHANGE

ISBN 978-981-12-8683-4 (hardcover)
ISBN 978-981-12-8739-8 (paperback)
ISBN 978-981-12-8706-0 (ebook for institutions)
ISBN 978-981-12-8707-7 (ebook for individuals)

For any available supplementary material, please visit
https://www.worldscientific.com/worldscibooks/10.1142/13693#t=suppl

Desk Editors: Gregory Lee/Amanda Yun

Typeset by Stallion Press
Email: enquiries@stallionpress.com

Contents

About the Author

M ichael Krause is a historian, author, and director of documentary films. He is especially interested in the people or institutions that are behind pioneering breakthroughs and who have contributed to a better understanding of the world and the universe. His publications on CERN, Nikola Tesla, German Angst, and the German Red Cross in the Third Reich have earned him worldwide recognition. His documentary film on climate change has become a standard work in German curricula. This book is based on his decades of research on the subject and serves as a guide to lead us safely through the unpredictable dangers of the future.

An Introduction: We Must Change

This book examines the financial, technical, and social situation of the world today and outlines some of the existing policies and technologies that will help us on our long and unpredictable journey through climate change. It tells the story of how this looming catastrophe came about, how humanity caused it, and how we, as individuals and communities, can meet the challenges ahead.

The climate crisis is the ultimate threat to human survival in the world today, and it can only be addressed through comprehensive technological and societal transformation. The main goal of the 2015 Paris Agreement was to limit global warming to below 2°C, with an ideal target of 1.5°C. To achieve this goal, we need to halve greenhouse gas emissions by 2030, halve them again by 2040, and then halve them again by 2050 to reach the net zero target ("Net Zero 2050"). This is necessary to avoid a climatic situation that can most aptly be described as an "ultimate catastrophe." This need for action raises the question of what to do with the 2,500 gigatons of carbon dioxide (CO_2) that humanity has released into the Earth's atmosphere since the beginning of the Industrial Revolution. As German Environment Minister Svenja Schulze put it, "The planet is in lethal danger, and with it its inhabitants." She said it is essential to prepare as best we can for the inevitable hazards of the climate crisis, such as disastrous flooding, relentless droughts, gigantic cyclones, and ever-larger wildfires.

The transition to Net Zero 2050 will require strong global social and political alliances. The appointment by German Foreign Minister Annalena Baerbock of former Greenpeace CEO Jennifer Morgan as her climate envoy is aligned with that type of collaboration. United States (US) President Joe Biden has declared that the world is in a climate crisis, and he has announced a climate goal of reducing emissions by 50–52% between 2005 and 2030. However, since the average US citizen has the largest carbon footprint in the world, a 50% reduction by 2030 would only bring the US's CO2 emissions down to existing levels in the European Union (EU). That said, the path to Net Zero in the US is more difficult than in most other countries, and many people in the US do not trust the administration to meet ambitious goals anyway. As for the broader problem of climate change, it is not much different for young people around the world. How can they trust their governments regarding measures against climate change when strong goals become very small steps along the process? COP26 — the 26th UN Climate Change Conference of the Parties in Glasgow in 2021 — has been heralded as the most important climate conference ever. The outcome following the UK-driven PR campaign and the 14-day conference was rather sobering. COP26 President and Minister of State in the UK Cabinet Office Alok Sharma put it this way: "The hope for 1.5 degrees remains alive, but the pulse is weak." Or, in the words of climate activist Greta Thunberg: "There was a lot of blah blah."

Power generation, transport, agriculture, and the food sector, as well as construction and the heating or cooling of buildings, must all become carbon-free as soon as possible. At the same time, the removal of CO2 from our atmosphere must be accelerated to reduce the risk of total ecological collapse on the planet. Three major factors that have led us to today's predicament — our economic system, our own inexcusable ignorance, and continued political inertia — must be addressed, discussed, and changed for the better. Such a transformation process will require significant financial investments, estimated at $5 trillion annually.

This book also deals with the concept of change. How is a successful change defined? What steps are necessary, and which steps come first? Where does change have to take place? How do we change ourselves and our different societies in practical ways? Importantly, this book does not provide a panacea, as tackling climate change is not a simple matter. The problem is not just climate change… it is ourselves. Our greed, our usury, and our stupidity are also to blame. In the course of the coming change, we will have to take responsibility for our previous actions because our own societal and all-encompassing irresponsibility brought us to this point. We have earned ourselves to death, and now we must give back to the Earth what we have taken from her.

Fortunately, there is good news. Innovation and human ingenuity will produce the tools needed to deal with the climate catastrophe. The only question is whether people will themselves be able to change to a sufficient degree. The whole process of change will take generations, and it will have to be a joint effort, and the stakes are incredibly high.

These are the components of a successful and sustainable climate policy:

1. Money… lots of money. Banks, large multinational corporations, and other global players are currently providing enormous sums of money to sufficiently address the climate crisis.
2. New technologies: Carbon capture and storage (CCS), direct air capture (DAC), and solar-to-fuel are transitional technologies for moving to a carbon-neutral energy economy. This book introduces, among others, two start-ups already operating worldwide, Synhelion and Climeworks, which produce carbon-neutral fuels from water, CO_2, and sunlight.
3. Many scientists are currently working to articulate the scientific and societal changes needed and how they can be integrated into our daily lives. For example, Professor Lorraine Whitmarsh focuses on how to achieve more environmentally friendly behavioral changes in society.
4. The younger generation, represented by the Fridays for Future movement, will hopefully play a major role in the transition to a

green world. Their tireless protest actions have made a significant difference: they have changed the global awareness of climate change. Who are these enthusiastic, politically active role models for the youth of the planet? What goals do they want to achieve, and what does their vision of change look like? What does the future look like in their eyes?

One thing is absolutely clear: We will all have to change in the future. This book is about the technical, personal, and cultural paths we must take to achieve this. We must utilize the climate shock as a catalyst to change ourselves and thus shape our own future.

The general issue is that over the last 120 years, humanity has doubled its production of goods or materials per generation (approximately every 20 years). The raw materials needed for such production are generally extracted from the Earth's biological cycle. During the same period, the non-human biomass on Earth has dramatically decreased due to the "reduction" of wildlife (whales, bison, fish stocks, etc.), global deforestation, destruction of soils and oceans, and the accelerated extinction of species.

This development has come to a critical point. Either we continue as before and "consume" the Earth, or we change our views, our goals, our politics, our economy, and ourselves — and do so radically and immediately. If we fail to change, the response or reaction of the biosphere will be fatal to untold living things on Earth, including massive numbers of humans.

How is it possible that people (politicians) who are considered wise ("sapiens"), and who have been given responsibility by us for our further coexistence, do not make the connection between radical exploitation and the associated damage the main subject of their politics? Why have we allowed this behavior to persist, and why have we all vigorously participated in it?

Humanity is now being given a second notice, as illustrated by these alarming trends. We are jeopardizing our future by not reining in our intense but geographically and demographically uneven material consumption and by not perceiving continued rapid population growth as a primary driver behind many ecological and even societal threats. By failing to adequately limit population growth, reassess the role of an economy rooted in growth, reduce greenhouse gases, incentivize renewable energy, protect habitats, restore ecosystems, curb pollution, halt defaunation, and constrain invasive alien species, humanity is not taking the urgent steps needed to safeguard our imperiled biosphere.

— "World Scientists' Warning to Humanity: A Second Notice,"
William J. Ripple and seven co-authors, 2017

The first item we need to develop further is our own resilience. We should not be overwhelmed by the multitude of new and unknown challenges but instead develop and maintain a stable, resilient state of mind. As former US President Franklin Delano Roosevelt said in the midst of the Great Depression, "We have nothing to fear but fear itself."

All the things we need to change, mitigate, and adapt for the future challenges of the climate crisis are already available. They are on the shelves of inventions that human ingenuity has produced. They are just waiting to be rediscovered, perhaps adapted, and then deployed. The biggest problem of all the big problems is humanity itself — our way of life, our behavior, our ways of thinking, our ego, and our unwillingness to adapt and change. It is time for all of us to leave our comfort zones. Either we win together or fail divided. We will decide whether climate change, climate crisis, climate catastrophe, or climate collapse will occur.

References

Climate Home News. https://www.climatechangenews.com/

European Commission. (2018). Action plan: Financing sustainable growth, March 8. https://eur-lex.europa.eu/legal-content/EN/TXT/PDF/?uri=CELEX:52018DC0097&from=EN

European Commission. (n.d.). Renewed sustainable finance strategy and implementation of the action plan on financing sustainable growth. https://ec.europa.eu/info/publications/sustainable-finance-renewed-strategy_en

Evans, S. (2021). Analysis: Which countries are historically responsible for climate change? Carbon Brief, October 5. https://www.carbonbrief.org/analysis-which-countries-are-historically-responsible-for-climate-change

International Institute for Sustainable Development. (n.d.). Sustainable Development Governance: Then and now. https://www.iisd.org/mission-and-goals/sustainable-development

United Nations. (n.d.). Goal 13: Take urgent action to combat climate change and its impacts. https://www.carbonbrief.org/analysis-which-countries-are-historically-responsible-for-climate-change/

I. Challenges and Revolutions

The Way to the Future

- It took Nature more than four billion years to create our modern society.
- Our modern society has taken 400 years to cause lasting damage to Nature.
- We need nothing less than a revolution to reverse this radically stupid development.
- Who actually came up with the idea that we could burn anything without worrying about the exhaust fumes that pollute the air we breathe?
- Who is responsible for this? In the end, it was us — me and you.

Revolutions are turning points in the history of mankind. Apart from the common understanding that revolutions are major upheavals that overturn order and that blood must be spilled, they also involve radical changes in the existing social, economic, and/or psychological order, often very rapidly. This transition can take a century, and at other times, it can take thousands of years, but sometimes it happens very quickly. In the history of mankind, there have been a great number of revolutions that have led to new forms of societies, but let us start at the beginning.

With fire.

The first hominids migrated from the cradle of humanity in Africa, which researchers have specifically identified as the "Rift Valley" in western Kenya. They were small groups of *Homo erectus* ("upright man"). The species evolved in the heart of Africa about 2 million years ago and could still be found in sites in Indonesia dating to 100,000 years BC. *Homo erectus* used handmade stone tools, such as the very practical Acheulean hand axe, for dissecting animal carcasses. These magnificent objects were used for cutting, scraping, and woodwork.

Stone axes were the first multipurpose tools of early man. Due to their extended period of use alongside constant evolution — they were in use for almost 1.5 million years — the Acheulean axe is the most sustainable technology ever developed by members of the species "Homo." It represents the evolution of complex behaviors expressed in the serial production of large tools with standardized shapes, suggesting advanced foresight and planning. The Acheulean Age is associated with a network of functional, economic, spatial, technological, and cognitive adaptations that drove human evolution. The cognitive and physical adaptations that supported the use of the axe may have also been related to climatic changes.

According to the "Savannahstan" hypothesis (Dennell and Roebroeks, 2005), groups of *Homo erectus* left their native Africa about 1.5 million years ago due to climate change at that time. There were no longer ice sheets over northern Europe, and excavations at the Israeli site of Ubeidiya, where these settlers probably arrived, show that the site was then inhabited by a variety of animals on which the new settlers from Africa could feed on. The *Homo erectus* was fast, agile, and intelligent enough to find its way in this new environment.

Rick Potts, director of the Smithsonian Institution's Human Origins Program, has been exploring how climate change is linked to evolutionary change. His theory is that during periods of rapid climate change, only individuals with certain traits survive, thrive, and raise children, who, in turn, can pass on these beneficial traits and thereby shape human evolution. For example, cognitive abilities that enabled humans to make sophisticated stone tools may have allowed their users to consume a wider variety of foods. Their curiosity may have led hominins to move to other climates as their original habitat dried up. *"Homo erectus* didn't have a map," Potts points out. "They didn't know they had left Africa. They just went to the next valley to see what was there." And from there, they kept moving.

Many important developments in hominins, such as the expansion of *Homo erectus* and *Homo sapiens*, coincided with periods of prolonged, severe climate change: presumably the emergence of bipedal australopithecines; the development of advanced stone tool technology in the African Rift Valley, the so-called "cradle of humankind"; the subsequent migration out of Africa; and the growth of the brain due to the challenges of survival in a foreign, hostile environment. How might climate change have shaped *Homo erectus*? Climate scientist Peter de Menocal, director of the Woods Hole Oceanographic Institution in Massachusetts, has been studying climate changes that occurred about two million years ago. He suggests that both *Homo erectus'* innate adaptability and rapid climate change may have led them to migrate out of Africa. Then, they became wildly successful because they were generalists and because they had developed a network of deep social relationships.

Did those early hominins regularly use fire? We don't know for sure, but we do know that they knew about fire, such as brush fires after a lightning strike or fires from volcanic activity. These natural fireplaces are quite apparent and must have been noticed by early humans. It seems fairly certain that very early Stone Age people (from about two million years BC) were already involved with fire in one way or another. So-called "fire harvesting," i.e., pouncing on prey fleeing from fire, is a common hunting method even among cheetahs. *Homo erectus* certainly did this as well. Initially, bushfires may have been a viable fire source for hominins before they incorporated it into their regular lives. The likelihood that these Stone Age hearths were preserved and could thus be discovered by researchers is extremely low. These remnants simply disappear, while stone tools like Acheulean axes naturally survive the test of time much better.

Figure 1. Acheuleen axe. (Public Domain)

All modern people need cooked food to survive, and the *Homo erectus*, as our natural predecessor, obviously did too. They had large brains, and those brains needed a lot of energy. Science teaches us that the regular use of fire evolved over time. Pyrotechnic abilities represent a significant change in hominid behavior, and they allowed for an expansion of food resources, protection from predators at night, and an improvement in social contact around a fireplace. The morphology of *Homo erectus* — with its larger body size, relatively smaller teeth, and likely less complex gut compared to its ancestors — suggests to some researchers that these hominids used controlled fire for cooking. At an excavation site in Koobi Fora, Kenya, the evidence of hearths has been dated to 1.5 million years BC. At Wonderwerk Cave in South Africa, excavations dating to about a million years BC indicate that massive amounts of grass and other vegetation were carried far into the cave and burned along with bones preserved as microscopic fragments. In the important site of Gesher Benot Ya'aqov in Israel, burned material (charcoal) has been preserved and dates to about 700,000 years BC.

The harnessing of fire was a long-term process that accompanied the further development of mankind. Whoever knew how to use it better represented an enormous advantage for the whole tribe and for it to thrive. It all began with fire.

References

Bar-Yosef, O. and Belmaker, M. (2017). Ubeidiya, in *Quaternary of the Levant Environments, Climate Change*, and Humans, Enzel, Y. and Bar-Yosef, O. (eds.). Harvard University, Massachusetts. https://www.cambridge.org/core/books/abs/quaternary-of-the-levant/ubeidiya/2BCEB616D3F5BBF26FE B03778747AD51

Dennell, R. and Roebroeks, W. (2015). An Asian perspective on early human dispersal from Africa. *Nature*, **438**:1099–1104.

Glausiusz, J. (2021). What drove homo erectus out of Africa? *Smithsonian Magazine*, October 19. https://www.smithsonianmag.com/science-nature/what-drove-homo-erectus-out-of-africa-180978881/

Hlubik, S., Cutts, R., Braun, D.R., *et al.* (2019). Hominin fire use in the Okote member at Koobi Fora, Kenya: New evidence for the old debate. *Journal of Human Evolution*, 133: 214–229. https://anthro.rutgers.edu/downloads/faculty/publications/1477-hominin-fire-use-2019/file

Institute of Geosciences and Earth Resources (n.d.). The Ethiopian Rift Valley. http://ethiopianrift.igg.cnr.it/cradle%20of%20mankind.html

Lamont-Doherty Earth Observatory (n.d.). Peter B. de Menocal. https://www.ldeo.columbia.edu/~peter/site/Home.html

Salazar, J. (2012). Climate change might drive human evolution. EarthSky, May 29. https://earthsky.org/human-world/peter-demenocal-climate-change-might-drive-human-evolution/

The Neolithic Revolution

The Neolithic Revolution, also known as the Agricultural Revolution, began about 12,000 years ago. For the first time, many people settled in houses, towns, and cities. They invented a set of rules with laws and bookkeeping, and they ate food grown on farms around their dwellings. All of this led to continued development and enough free time to learn, research, and invent. This change was tremendous, with enough food and (more or less) friendly people in a (more or less) friendly society. In short, the Neolithic Revolution led directly to the life we lead today.

The cradle of modern civilization stood in the Fertile Crescent, a crescent-shaped region in the Middle East stretching from modern-day Israel to the Persian Gulf. Agriculture, irrigation, and the wheel were developed in this area. The region has always been characterized by migrants moving from Africa to areas of the steppe and Central Asia. The central region of the Fertile Crescent is Mesopotamia, from the Greek for "between the rivers," i.e., between the Tigris and Euphrates rivers in present-day Iraq.

In the Jordanian desert, near the present-day capital of Amman, the first pre-Ceramic sites of the Neolithic period have been excavated, e.g., the 9,000-year-old farming settlement of 'Ain Ghazal (source of the gazelles, dated to ca. 6750 +/− 80 BC). The oldest humanoid sculptures and death masks to date were found here. The heads of the statues are colored with red ocher, and the inlaid eyes look out with reason and intelligence, but also with kindness, as if they were created to escort their living souls into eternity. These relics are clear evidence of the artistic skill of the people who created them and an indication that they had an idea of what we call the soul.

Catal Hüyük was a relatively large proto-city in present-day Turkey that was inhabited from about 7500 to 6400 BC. The population has been estimated to have been up to 10,000 people. Dwellings were arranged in a honeycomb-like maze and were accessible through holes in the ceiling and doors on the sides of the houses. The roofs served as streets. The "Seated Woman of Catal Hüyük" represents a corpulent goddess, a type of figure found at many Middle Eastern excavation sites from this period. The large number of statues found has led some scholars to theorize about a general belief in an all-encompassing mother goddess, Gaia (theory of Mother Earth), and the like. This led to modern theories of goddesses worshiped by peaceful, matriarchal, agrarian societies. This prehistoric matriarchal theory ("Magna Mater") as proof of a peaceful matriarchy is regarded as unlikely in modern academic circles, although the sheer number of women depicted, coupled with the absence of male figures that could be regarded as figures of a god (or male ruler), is certainly significant.

Figure 2.   "Mother Goddess," Catal Hüyük, Museum of Anatolian Civilizations.

The Bronze Age Revolution

During the Bronze Age (approximately 3300 BC to 1200 BC), new forms of income (trade) and, thus, prosperity lifted the prehistoric age of mankind to a new level. By definition, prehistory is characterized by the absence of writing, but at the beginning of this age, proto-writing with a few letters and numbers for trade figures relevant to accounting was already emerging. Early practical writing systems were then developed in Mesopotamia and Egypt; consequently, powerful dynasties arose throughout the Fertile Crescent and lasted for the next millennium. Egypt began to take a dominant role, and Mesopotamia flourished, with great cities whose names still resonate today, such as Babylon and Ur.

The Bronze Age was a period when metals were used extensively and large trade networks developed. Bronze, an alloy of tin and copper, is harder and more durable than other metals. Tin was mined in Britain and had to be transported to bronze production sites in Greece, the Middle East (Timna Valley, Israel), and other places.

During the Bronze Age, conflict management in large parts of ancient Europe underwent a significant change. Metal became an everyday commodity that people could use for violent action. Swords, spears, shields, and armor were developed for the battlefield. This was perhaps the most profound change in modern human history. The mechanisms of warfare (instruments, tactics) invented during that time remained in use for the next millennia — and to this day. Metal technology made new forms of objects possible, and from then on, advanced and specialized weapons were produced for the class of warriors at the top of society.

Spears and shields became "peacemakers" of the time. Shields were used to form a wide defensive front for battle lines with groups of warriors. Spears were the main offensive weapons, while helmets and other pieces of armor were used to protect the warriors' bodies. Scenes showing how these items were used in another pastime, lion hunting, are depicted on metal relics found in the shaft tombs of Mycenae (17th to 15th centuries BC) and elsewhere. Mycenae was a hilltop fortress that could repel attackers of any kind. At the same time, the sheer size and appearance of the complex represented power. The main entrance to the citadel of Mycenae, called the "Lion's Gate" in modern terminology, is the largest surviving sculpture in prehistoric Greece. The gate remains impressive and majestic, a symbol of the proud self-confidence of its builders.

Figure 3.   Lion Gate, Mycenae. (Public Domain)

Making use of the new weapons required skilled and trained warriors, skilled craftsmen, and a considerable investment from the entire proto-state. Warriors had become an important part of society, wherein fighters could even be described as a kind of "corporate identity" of the Bronze Age.

The Bronze Age presented a new facet of human society — war. Moreover, the new technologies — the production of weapons and armor made of metal — became an engine of economic development. The new technologies required new forms of trade throughout Europe. The Bronze Age is characterized by a vast trade network, brilliant works of art and artifacts, early writing and record keeping, and the standardization of language. Akkadian was used throughout Babylonia, Akkad, and Assyria for official documents and as a spoken language.

The Epic of Gilgamesh is the oldest Heroic Epic in the world. It tells the story of the legendary king Gilgamesh (about 2700 BC), who is two-thirds god and one-third man. Gilgamesh is a despot. Therefore, the goddess Aruru decides to create the steppe dweller Enkidu to defeat Gilgamesh. A battle ensues, which ends in a draw. Gilgamesh and Enkidu become friends and share adventures. This epic from Mesopotamia is about power, friendship, transformation, and the longing to overcome death. The most important themes of humanity were spelled out in this early work.

References

Barjamovic, G., Chaney, T., Coşar, K., and Hortaçsu, A. (2019). Trade, merchants, and the lost cities of the Bronze Age. *The Quarterly Journal of Economics*, 134(3): 1455–1503. https://academic.oup.com/qje/article/134/3/1455/5420484

Mineralienatlas — Fossilienatlas. https://www.mineralienatlas.de/lexikon/index.php/Mineralien-portrait/Kupfer/Kupfer%20im%20Chalkolithikum%20%28ca.%205.500%20-%202.200%20v.%20Chr.%29?lang=en&language=english (in German)

The William R. and Clarice V. Spurlock Museum of World Cultures. (n.d.). Trading in the Bronze Age. https://academic.oup.com/qje/article/134/3/1455/5420484

The Christian Revolution

D uring the Christian Revolution, a Chaldean named Jesus of Nazareth was crucified by the Romans. At that time, Rome was the epicenter of the world, and crucifixion was the most disgraceful fate imaginable throughout the empire. It was a punishment for slaves. But after this Chaldean's crucifixion, Christianity irreparably changed the world and continues to do so today. The backbone of this revolution became the Christian Bible, in which the entire history of mankind is contained in all its aspects. At the center of this new form of faith was the death of God, or rather, the death of the Son of God made man. God remained in his (or her) sphere and continued to decide the fate of every human being, dead or alive. A tricky construct, but solid and, through constant repetition, something in which everyone could believe.

Christianity bridged the gap between Jewish thought, culture, and faith and the pagan religion of the Roman world. Christianity offered non-Jews the opportunity to enter into a full relationship with the God of Abraham, Isaac, and Jacob. Christianity retained the Jewish understanding of God as the only Creator of the world (Who gives free will to all people) while eliminating the earlier worship of unearthly creatures as gods. Christianity adopted some of the old myths and rites but also created new ones, such as the idea that God actually spoke to people like Abraham or Moses. The immaculate conception of Mary and the resurrection of Jesus were presented as true events that had occurred at a specific time and place and had been confirmed by several eyewitnesses. Christianity contained three ideas, in particular, that would be highly influential in the further development of human culture.

The first idea entails the notion that God is rational and of a creative nature. John 1:1 says, "In the beginning was the Word, and the Word was with God, and the Word was God." The "Word" here is the translation of the Greek "logos," which may mean "reason," "speech," or "word." God is NOT depicted herein as irrational but full of truth and intelligence. This positions Christ as the incarnation of reason, which opened the possibility for normal people to be and act rationally.

This highlights the second point that Christianity emphasizes — the assertion that all people are capable of truth through reason. Consequently, all people in the world were allowed to know basic truths about right, wrong, good, and evil through their own rational thinking. At the same time, this placed all people on equal footing with one another. For a long time, this Christian ethic of "doing good on the ground of truth" stood at the center of human society.

Figure 4. Jesus gives the Great Commission to the disciples, Master of the Reichenau school of illumination. (Public Domain)

From that moment (the inception of human reasoning), people could decide to act rightly. This also means that they could act differently, i.e., wrongly. The basic idea behind all of this is the freedom of choice, the third great idea that Christianity added to human history. Christ himself had told his followers that they were free to follow him or not. This kind of freedom also means that there is a limit to the ability of others — including the government or the state — to tell people what to do. This kind of intellectual freedom — of being able to decide what to do — also implies the freedom to use our minds to understand the world. This is the basic idea of research and man's drive to know everything that was previously unknown.

The conquest of the world and the subjugation of all things on this earth under the rule of man is a cornerstone of Christian doctrine. Genesis 1:28 says, "And God blessed them, and said unto them, Be fruitful, and multiply, and replenish the earth, and subdue it; and have dominion over the fish of the sea, and over the fowl of the air, and over every creeping thing that creepeth upon the earth." Some synonyms for "subdue" include "conquer," "oppress," and "subjugate." This ideology of "subduing the earth" is one of the basic tenets in the ideology of the Christian church, even though some Christian scholars today deny that this is the case. This denial after two millennia of Christian history is whitewashing and evasion. The doctrine of conquest remains one of the deepest layers of Christian ideology. In 1928, John Widtsoe, then director of the US Federal Bureau of Reclamation, wrote: "The destiny of man is to possess the whole earth; and the destiny of the earth is to be subject to man."

We must free ourselves from this way of thinking, for we are a part of the Earth, not its owner.

References

Encyclopedia Britannica. (n.d.). History of early Christianity. https://www.britannica.com/topic/history-of-early-Christianity

Gregg, S. (2019). The Christian revolution. *Religion & Liberty*, **29**(2). https://www.acton.org/religion-liberty/volume-29-number-2/christian-revolution

History. (2017). Christianity, October 13. https://www.history.com/topics/religion/history-of-christianity

Wikipedia. (n.d.). Christianity. https://en.wikipedia.org/wiki/Christianity

Wikipedia. (n.d.). Early Christianity. https://en.wikipedia.org/wiki/Early_Christianity

Wikipedia. (n.d.). Sola scriptura. https://en.wikipedia.org/wiki/Sola_scriptura

The Renaissance Revolution

A lthough fiercely disputed among historians, the era designated by the term "Renaissance" (rebirth) has all the ingredients of a revolution because of the period's far-reaching changes in art, politics, science and technology, and religion. The bloody part of that particular revolution was more or less transferred to newly discovered areas in America and the Far East. After what was then (inaccurately) called the "Dark Ages," the Middle Ages became a time of change, and "The Garden of Earthly Delights" (c. 1490–1510), a painting by Hieronymus Bosch (c. 1450–1516), can be considered central to this shift. Explicitly and intentionally strange, the painting can be interpreted as a warning about the dangers of temptation, or as a lush panorama of carnal desires. Primarily, it is a glimpse of the inner thoughts of a brilliant mind interpreting the world as he sees it. The painting is an interpretation of time — an allegory and a mystery at the same time.

Figure 5.   The Garden of Earthly Delights, Hieronymus Bosch (c. 1450–1516). (Source: Hieronymus Bosch, http://boschproject.org/dzi/00MCPVIS.dzi)

One of the main drivers of the Renaissance was the emergence of competition between cities, merchant houses, and families, especially in Northern Italy and the Netherlands, which were then governed by the Dukes of Burgundy. Florence, for example, was ruled by the Medici, a powerful clan that patronized artists and earned the money to do so with its own "Banco Medici." Terms widely used today, such as account, giro, credit, or bankruptcy, originate from that era of proto-capitalism. Florence developed into a central hub for money and trade, and its wealthy patrons could afford to import insanely expensive paintings from Flanders (e.g., van Eyck), thus giving impetus to the artists of Northern Italy.

Renaissance art emerged as an independent style in Italy around 1400 AD. Since the art of classical antiquity was considered the noblest of all ancient traditions, Renaissance art was characterized by realism and naturalism. Renaissance artists such as Flemish Jan van Eyck (c. 1390–1441), the inventor of oil painting, and the Italian Titian (c. 1488–1576) had a strong interest and awareness of nature, learning, and individualism, and they wanted to depict people and objects in a naturalistic way.

The Renaissance was a time of brilliant minds such as the Italian polymath Leonardo da Vinci (1452–1519), who painted the world-famous "Mona Lisa" and "The Last Supper." Other painters from the region are Sandro Botticelli (1445–1510), Michelangelo (1475–1564), Titian, and Raphael (1483–1520). The invention of printing by the German Johannes Gutenberg (c. 1400–1468) in 1450 enabled better communication throughout Europe and faster dissemination of new ideas and scientific inventions.

Exploration of the world opened new countries and civilizations to European trade. In this age of discovery, European adventurers, explorers, and entrepreneurs launched myriad expeditions to travel the globe. Some discovered new shipping routes to America, India, and the Far East — Christopher Columbus, Marco Polo, and Amerigo Vespucci, respectively. Their risky ventures proved to be highly lucrative for their kings and employers and a complete disaster for the indigenous populations of the new territories they discovered, especially in America. In a very short amount of time, up to 90% of the native population in the Americas was eradicated, mainly from infectious diseases contracted through contact with Europeans.

During the scientific renaissance, knowledge of nature from antiquity was rediscovered, such as by the Polish polymath Nicolaus Copernicus (1473–1543), who placed the sun at the center of the solar system — a theory that had already been developed by the Greek astronomer Aristarchus of Samos (c. 310 – c. 230 BC) as well as by the Swiss-born Paracelsus (c. 1493–1541), a pioneer of the "medical revolution" of the German Renaissance.

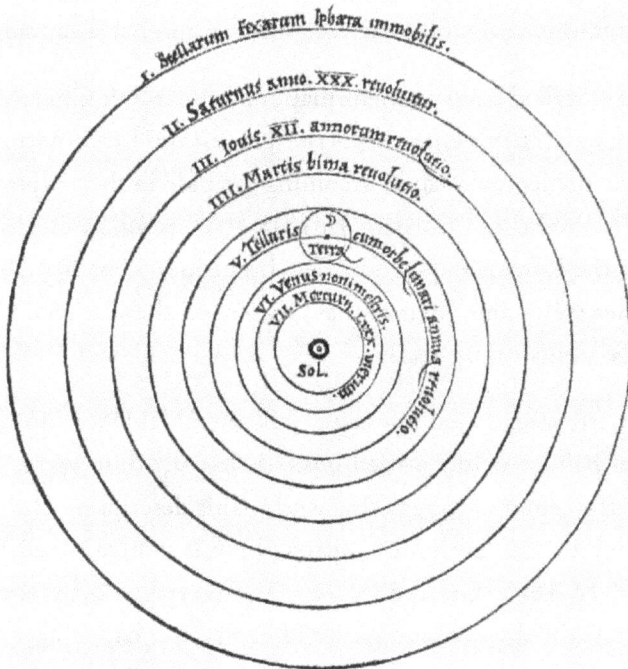

Figure 6.   Heliocentric model, Nicolaus Copernicus, 1543. (Public Domain)

The decline of the Renaissance was the result of several converging factors. Between 1494 and 1559, the Italian peninsula was ravaged by the Italian Wars, also known as the Habsburg–Valois Wars. Changing trade routes led to economic decline in Italy, thus limiting the means of wealthy investors.

More and more people had learned to read, write, and interpret things by the end of the Renaissance. As a result, they began to closely examine and criticize religion as they knew it. In addition, printing made it possible to easily reproduce and widely distribute new texts and pamphlets, including the Bible. Martin Luther (1483–1546), a German Augustinian friar and professor of theology, published a pamphlet in 1517 — known as the 95 Theses — against the sale of indulgences and other shortcomings of the existing Catholic Church. The movement he triggered affected all sectors of society and eventually founded a new denomination, Protestantism.

During the so-called Counter-Reformation, the Catholic Church censored artists and writers in response to the Protestant Reformation. Many Renaissance intellectuals feared becoming too bold in their expression and art, which limited their creativity. In 1545, the Catholic Council of Trent introduced the Roman Inquisition. The Inquisition classified Humanism and all other views and dogmas that challenged the Catholic Church as heresy, to be punished by death.

The Italian astronomer and physicist Galileo Galilei (1564–1642) was an advocate of heliocentrism, which posited that the Sun, rather than the Earth, was the center of the existing world. This description of the cosmos ran contrary to the Catholic Church since it contradicted the Holy Scriptures. In 1615, Galileo's case was investigated by the Roman Inquisition. Galileo was found guilty of heresy and had to spend the rest of his life under house arrest. The Church had already played a powerful role as a drag on innovation during the "Dark Ages," from the fall of the

Roman Empire to the beginning of the Renaissance. It retained its power just as strongly and relentlessly into our times, but reason and science were to become its strongest enemies.

References

Pirie, M. (2019). Leonardo da Vinci — Polymath. Adam Smith Institute, April 19. https://www.adamsmith.org/blog/leonardo-da-vinci-polymath

Who's Who: The People Lexicon. (n.d.). Leonardo da Vinci. https://whoswho.de/bio/leonardo-da-vinci.html

Wikiart. (n.d.). Jan van Eyck. https://www.wikiart.org/de/jan-van-eyck

Wikipedia. (n.d.). Science in the Renaissance. https://en.wikipedia.org/wiki/Science_in_the_Renaissance

Wikipedia. (n.d.). The Garden of Earthly Delights. https://en.wikipedia.org/wiki/The_Garden_of_Earthly_Delights

The Revolution of the Enlightenment

I n the 16 hundred years that followed the crucifixion of Jesus, Christian ethics and the precepts laid down in the Bible had time to become the fundamental ideology of Western civilization. After the cultural and intellectual upsurge brought about by the scientific revolution of the Renaissance and the subsequent critique of the prevailing dogmas of the Church, it was time for a reversal. Although a host of scholars, scientists, writers, and other bright minds recognized that the Catholic myth of God was, in their eyes, an elaborate hoax, it was to endure for quite some time. However, with scholars and scientists such as Isaac Newton, Francis Bacon, Rene Descartes ("I think, therefore I am"), David Hume, Gottfried Wilhelm Leibniz, Adam Smith, John Locke, and Immanuel Kant ("Have the courage to use your own intelligence!") — to name but a few — the Age of Enlightenment was to be led by these grandiose thinkers toward new objectives of human history.

Niccolò di Bernardo dei Machiavelli (1469–1527), the father of modern political theory, is one of the most influential and well-known pioneers of the Enlightenment. He served as a diplomat in the Italian Republic of Florence during the Medici family's 14-year exile before writing the book *The Prince*, for which he is still famous today. The book was intended as a handbook for unscrupulous politicians, and it is full of mean tricks and maneuvers. *The Prince* was so innovative because it emphasized practical and pragmatic strategies rather than philosophical ideals. Machiavelli's treatise did not bring him the success he sought, namely, a much-desired political office. His cool, rational mind was rediscovered as one of the most important sources of Enlightenment thought (though only after his rediscovery and rehabilitation) with great enthusiasm by the educated class at the beginning of the 19th century.

Francis Bacon (1561–1626) paved the way for the Enlightenment. He became the prophet of the coming scientific revolution despite never having been a scientist. After a career as a lawyer, he began writing about the necessary renewal of the human mind: "The true and lawful goal of the sciences is none other than this: that human life be endowed with new discoveries and power." Isaac Newton (1642–1727) was the greatest scientist of the Age of Enlightenment, or at least the best known. His method was based on methodical questioning — focusing on what could be known — and the relentless observation of Nature. Newton developed a rudimentary version of infinitesimal calculus, discovered the law of optics (white is a composite color), and established the fundamental principle of gravitation in his seminal work *The Mathematical Principles of Natural Philosophy*, published in 1687, which he derived from Kepler's *Laws of Planetary Motions* (published 1609–1619). Newton's work is one of the most important cornerstones of physics and opened the door to further research beyond the boundaries of the Earth, marking the peak of the Age of Enlightenment.

At the heart of the Enlightenment was the quest for greater knowledge about human rationality and man's situation in society. This "Age of Reason" was about freeing society from the shackles of the past centuries. New ideals such as scientific progress, human freedom, separation of church and state, and the search for new forms of government characterized this prosperous era. These new ideas were discussed extensively in "intellectual" circles such as academies, salons, and Masonic lodges. Eventually, the quest for individual freedom and tolerance met resistance from the Catholic Church and challenged the political rule of the absolutist monarchy. The movement became dangerous, especially for those in power in France where, during the French Revolution, the king and queen were beheaded, as were thousands of people who were considered supporters of the old regime or not radical enough in the eyes of the revolutionaries.

Figure 7.   The Storming of the Bastille, Henry Singleton (1766–1839). (Public Domain)

The French Marquis de Condorcet (1743–1794) formulated the goal of this period: "The time will therefore come when the sun will shine only on free men who know no other master but their reason." Moses Mendelssohn (1729–1786, German-Jewish philosopher) put it differently when he described the Enlightenment as a "yet uncompleted process of education in the use of reason, which should be open to all," calling the movement to spread Enlightenment ideas among the lower social classes. The Prussian philosopher Immanuel Kant (1724–1804) was well aware that the unrestricted development of reason, if it went too far and challenged or redefined existing meanings without limit, could throw the social, religious, and political order into chaos. Condorcet himself had to flee Paris after his leading role during the French Revolution (1789–1799) turned his more radical opponents against him. He had opposed the execution of the king, and in March 1794, Condorcet was found dead in his prison cell. All revolutions devour their own children.

Immanuel Kant developed a positive view toward the further development of reason when he defined the Enlightenment as "the liberation of man from his self-induced immaturity." Man was now free to think for himself, without guidance from others (e.g., the church or governments). Today, we understand the far-reaching consequences of the Enlightenment more globally. It was a movement in which intellectual projects transformed society, science, behavior, and governments, not only in France but throughout the world. It moved people to think radically, stand on the shoulders of their predecessors, use their minds and reason, and have courage. This may sound utopian and dangerous to some, but it remains the only way out of the apparent dead ends of human development.

Peter Gay (1923–2015, German-British historian) considered the Enlightenment to be an anti-religious movement that strove for freedom and progress and was driven by a justified urge to change man's relationship with himself and society. Gay even linked the Enlightenment to the founding of the American colonies from England and the Republic of the United States (US) that followed. He argued that the American Declaration of Independence of July 4, 1776, and its commitment to "Life, Liberty, and the Pursuit of Happiness" were the fulfillment of the Enlightenment's ambitious goals. From this point of view, the Enlightenment had as great an impact on the world as Christianity.

References

Britannica Encyclopaedia. (n.d.). Enlightenment. https://www.britannica.com/event/Enlightenment-European-history

Britannica Encyclopaedia. (n.d.). Immanuel Kant. https://www.britannica.com/biography/Immanuel-Kant

Kant British Library. (n.d.). The Enlightenment. https://www.bl.uk/restoration-18th-century-literature/articles/the-enlightenment

Study Smarter. (n.d.). Age of Enlightenment. https://www.studysmarter.co.uk/explanations/english-literature/literary-movements/age-of-enlightenment/

The National Archives. (n.d.). French Revolution. https://www.nationalarchives.gov.uk/education/resources/french-revolution/

Wikipedia. (n.d.). Francis Bacon. https://en.wikipedia.org/wiki/Francis_Bacon

Wikipedia. (n.d.). Immanuel Kant. https://en.wikipedia.org/wiki/Immanuel_Kant

Wikipedia. (n.d.). Niccolò Machiavelli. https://en.wikipedia.org/wiki/Niccol%C3%B2_Machiavelli

World History Encyclopedia. (n.d.). French Revolution. https://www.worldhistory.org/French_Revolution/

The Industrial Revolution

The Age of Enlightenment came to a bloody end in France at the end of the 18th century. King Louis XVI was publicly beheaded by guillotine at the Place de la Revolution in Paris on January 21, 1793. In Great Britain, a new era was already beginning. The Industrial Revolution brought about the transition from agricultural or artisanal production to a new economy dominated by machines. New manufacturing and production processes, the dramatically improved use of water power, the increasing use of steam power, and the development of complex machine tools became the driving forces behind the rapid growth of the British economy.

Figure 8.   Execution of Louis XVI — copperplate engraving, 1793. (Public Domain)

The iron and textile industries played a central role in the Industrial Revolution, and the steam engine of James Watt (1736–1819) was the key innovation behind them. In the 1760s, Watt, a Scotsman, had improved a machine designed by Thomas Newcomen so that it could therefore produce rotary motion. Additionally, Watt's new steam engine was much cheaper to make. The steam engine was used in the development of textiles, mining, railroads, shipping, steel production, and many other industries. The spinning machine, coke smelting, and the puddling and rolling process for iron production were other inventions and improvements that rapidly spurred British economic growth. The new processes brought about larger workshops that required more workers to labor longer hours. The rise of these new industries created a new infrastructure that led to the emergence of new types of industrial cities such as Manchester and Birmingham. The Industrial Revolution required a lot of cheap labor, and with it, some people (inventors, investors) got rich quickly, widening the already existing gap between the rich and the poor.

Figure 9.   Coal pits in the black country, West Midlands, England. (Public Domain)

The British colonies became the main source of critical raw materials such as cotton, sugar, and tobacco. British coal, mined by purposely underpaid British and Welsh workers — many of them children — became the main source of energy for the novel machines. The British wanted to maintain their monopoly for as long as possible, so they banned the export of their revolutionary machines. It took more than 50 years before the first workshops were built on the continent — in Belgium. During the monopoly (c. 1750–1810), England became the richest country in the world.

Extensive investments in infrastructure such as roads, bridges, canals, port facilities, and railroads enabled England to develop faster and further than any other country. On the continent, similar industrial areas first emerged gradually in northeastern France, Germany (especially the Ruhr and Silesia regions), and, later, the northeastern parts of the US. The growing industrial society, with its new patterns of settlement, fundamentally changed the social structure of the affected communities and populations. The contrast between the haves and have-nots and the resulting social question of wealth inequality led to a critical discourse on the emerging social forms and the prevailing culture of the powerful and the rich.

Adam Smith (1723–1790), the Scottish philosopher and pioneer of political economy, was the first to explain the modern economy in terms of specialization, division of labor, market transactions, and increased productivity. In 1776, the same year that Watt introduced his steam engine, Smith published his *magnum opus*, *The Wealth of Nations*. Smith argued for free enterprise (i.e., free from government intervention) and private ownership of the means of production. Adam Smith applied the political formula *divide et impera* ("divide and rule") to the modern capitalist world: "The greatest improvement in the productive powers of labour ... seem[s] to have been the effects of the division of labour."

In 1752, Smith, who was characterized by contemporaries and friends such as David Hume as a typical "absent-minded professor," was appointed as professor of moral philosophy at the University of Glasgow. There, he lectured on "ethics," "natural theology," jurisprudence, and political economy. Smith's main subject was what we now call "social science." In his writings, Smith developed the concept of our modern society, and man as a part of that society.

Adam Smith held the view that society benefits most when everyone pursues their own interests: "Every individual necessarily labours to render the annual revenue of the society as great as he can." In this way, Smith envisioned, an "invisible hand" of the free market would transform the individual pursuit of profit into a general benefit for society. According to Smith, this automatism was to take place only within domestic (British) industry and was not intended to apply to foreign markets or to trade between domestic and foreign markets.

Today, many scholars and scientists invoke Adam Smith's arguments to defend the moral necessity of the excessive pursuit of self-interest. Smith did not say or mean anything like that, but it reveals how Smith's vision can be exploited when necessary. Smith himself was rather critical of the true interests of the merchants and manufacturers of his era. In his opinion, they understandably tended to deceive and oppress people and society by pursuing only their own interests and nothing else. Essentially, Smith saw man as a social being and viewed compassion as the core of moral judgment and social interaction.

In reality, the doctrine of free trade and the free market allowed factory owners to dictate the working conditions themselves. Workers' real wages did not increase during the Industrial Revolution, which resulted in factory workers becoming members of the "lumpenproletariat" (according to Karl Marx). Industrialization brought unprecedented innovation, growth, and prosperity, but it also meant increasing inequality, poverty, misery, and sometimes death. The system that later came to be called "capitalism" now began to unfold in full force. In *The Wealth of Nations*, Adam Smith explains the capitalist system's core functionality: "When an independent workman, such as a weaver or shoemaker, has got more stock than what is sufficient to purchase the materials of his own work, and to maintain himself till he can dispose of it, he naturally employs one or more journeymen with the surplus in order to make a profit by their work. Increase this surplus, and he will naturally increase the number of his journeymen. The demand for those who live by wages, therefore, necessarily increases with the increase of the revenue and stock of every country, and cannot possibly increase without it. The increase of revenue and stock is the increase of national wealth. The demand for those who live by wages, therefore, naturally increases with the increase of national wealth, and cannot possibly increase without it."

The enormous growth of capitalism during the Industrial Revolution was based on the use of accumulated capital to constantly expand production capacity. The policies of 19th-century liberalism meant free trade, sound money (based on the gold standard, by which paper money could be fully exchanged for gold at any time), balanced budgets, and an absolute minimum of social benefits (poor relief). The purest form of capitalism is free market or *laissez-faire* capitalism ("Manchester Liberalism"), in which individuals are fully free from government supervision and constraints. They can decide for themselves where and when to invest, what to produce or sell, and at what prices to exchange goods and services. The *laissez-faire* market operates without control or supervision. This purest form of capitalism is still the hallmark of modern liberal economists.

A Christmas Carol, the novella famously written by Charles Dickens (1812–1870), was published in 1843. The main character of the book is Ebenezer Scrooge, an arch-capitalist who has his assistant working in a bitterly cold room, and when the assistant asks his employer if he could go home early on Christmas Eve for once, Scrooge responds with his iconic expression, "Bah, humbug!" In the book, he is presented as the typical capitalist of the Victorian age. He is stingy and ruthless, and he will stop at nothing to increase his profits by squeezing more work out of his employee. Dickens, who had lived in great poverty himself, could not imagine a passionate, perhaps even revolutionary, story centered around the capitalist Scrooge. Instead, the author chose to write an idealized fairy-tale transformation of his main character. In the end, Scrooge finally becomes a "nice" capitalist. In some ways, this could act as a model for our own transitional times so that we might finally become kinder to our own world.

References

Encyclopaedia Britannica. (n.d.). Mercantilism. https://www.britannica.com/money/topic/mercantilism

Jahan, S. and Mahmud, A.S. (n.d.) What is capitalism? International Monetary Funds. https://www.imf.org/en/Publications/fandd/issues/Series/Back-to-Basics/Capitalism

Kaczynski, T. (1995). Industrial society and its future. http://editions-hache.com/essais/pdf/kaczynski2.pdf

Mohajan, H. (2019). The First Industrial Revolution: Creation of a new global human era, MPRA Paper No. 96644, May 30. https://mpra.ub.uni-muenchen.de/96644/1/MPRA

Smith, A. (n.d.). *Book 1: On the Causes of Improvement in the Productive Powers. On Labour, and on the Order According to Which its' Produce is Naturally Distributed Among the Different Ranks of the People.* https://www.marxists.org/reference/archive/smith-adam/works/wealth-of-nations/book01/ch01.htm

The Victorian Web. (n.d.). On "the invisible hand" — Adam Smith. https://victorianweb.org/economics/smith3.html

Wikipedia. (n.d.). A Christmas Carol. https://en.wikipedia.org/wiki/A_Christmas_Carol

Wikipedia. (n.d.). Manchester Liberalism. https://en.wikipedia.org/wiki/Manchester_Liberalism

The Industrial Revolution 2.0

T he end of the 19th century was characterized by an abundance of inventions and technologies, some of which were completely new (e.g., the telegraph and electricity). This enabled rapid economic growth and the accumulation of immense wealth. The book *The Gilded Age: A Tale of Today* by American writer Mark Twain (1873, co-authored with Charles D. Warner) gave this age its nickname. Never was there a time of faster growth in the US than in the 30 years between 1870 and 1900. The construction of railroads was the most important factor in this surge of growth, as it enabled the rapid and safe transcontinental transportation of goods and people. From then on, wherever labor was cheap, factories could be built to produce goods for national distribution. The development of railroads and the resulting growth of industry helped the US rise to the top of the global economy. European capital from Victorian Britain, Belle Epoque France, and Germany poured into the burgeoning American financial system, centered on Wall Street. This capital flowing into the US was invested primarily in heavy industry; soon, American steel production exceeded that of Britain, Germany, and France combined.

During the "Gilded Age," or rather, the Industrial Revolution 2.0, the US transformed from a pre-industrial and rural nation to one of the world's leading industrialized nations. From 1860 to 1890, 500,000 patents were issued in the US. American Thomas Alva Edison (1847–1931) invented almost everything in the field of applied technology — the phonograph, the motion picture camera, and the electric light bulb, to name just a few. His system of electricity generation and distribution, the direct current (DC) system, was the first to provide electric current in a limited range. However, his DC system soon faced serious competition. The alternating current (AC) system invented by Nikola Tesla was clearly superior to Edison's DC system. The new system (AC) was introduced by George Westinghouse. This led to a rapid expansion of electricity supply in cities in the US, but also in France, Great Britain, and Germany. Soon, electricity was available almost everywhere.

Figure 10.   Thomas Edison with his second phonograph, photographed by Levin Corbin Handy in Washington, April 1878. (Public Domain)

After the railroad and the power grid, the next big thing to boom was the oil industry, which became the engine of the world economy. This development brought further growth and even more prosperity. That said, the rapidly growing use of fossil fuels eventually led to the catastrophic state of our environment today. At the end of the "Gilded Age" around 1900, John D. Rockefeller and his Standard Oil Company controlled 90% of the American oil business. The company initially produced mainly kerosene for the oil lamps that were still widely used at the time. It was so cheap that competing products, such as whale oil, coal oil, and, for a time, electricity, lost out in the race to light American homes, factories, and streets.

One of the side effects of rapid economic transformation in the US was that most Americans were living better and healthier lives. Life expectancy had increased, largely because of a dramatic decline in infant mortality. However, with economic progress came new waves of poverty and inequality, and the gap between the rich and the poor continued to widen.

The Industrial Revolution 2.0 led to extreme poverty, especially among the American rural population, countered by extreme wealth among entrepreneurs. Successful industrialists were often referred to as "robber barons" because their business methods were considered unethical or unscrupulous. Henry Ford, Andrew Carnegie, Cornelius Vanderbilt, and John D. Rockefeller are famous business magnates who were accused of being such robber barons — monopolists who made profits by, for example, deliberately limiting the production of goods and then raising prices.

Andrew Carnegie (1835–1919) made his giant fortune in the steel business, later becoming one of the most important philanthropists of the age. In his book *The Gospel of Wealth*, Carnegie wrote: "The price which society pays for the law of competition, like the price it pays for cheap comforts and luxuries, is also great; but the advantage of this law is also greater still, for it is to this law that we owe our wonderful material development, which brings improved conditions in its train. But, whether the law be benign or not, we must say of it, as we say of the change in the conditions of men to which we have referred: It is here; we cannot evade it; no substitutes for it have been found; and while the law may be sometimes hard for the individual, it is best for the race, because it insures the survival of the fittest in every department."

References

Encyclopaedia Britannica. (n.d.). Germany: The economy, 1890–1914. https://www.britannica.com/place/Germany/The-economy-1890-1914

Encyclopaedia Britannica. (n.d.). Nikola Tesla. https://www.britannica.com/biography/Nikola-Tesla

Insider. (n.d.). Cornelius Vanderbilt. https://www.businessinsider.com/robber-barons-who-built-and-ruled-america-2017-7#cornelius-vanderbilt-dominated-the-steamship-business-in-long-island-sound-and-built-an-empire-of-railroads-around-new-york-city-2

Library of Congress. (n.d.). U.S. history primary source timeline. https://www.loc.gov/classroom-materials/united-states-history-primary-source-timeline/rise-of-industrial-america-1876-1900/overview/

Wikipedia. (n.d.). Andrew Carnegie. https://en.wikipedia.org/wiki/Andrew_Carnegie

Wikipedia. (n.d.). Category: 20th century revolutions. https://en.wikipedia.org/wiki/Category:20th-century_revolutions

Wikipedia. (n.d.). Mark Twain. https://en.wikipedia.org/wiki/Mark_Twain

Wikipedia. (n.d.). Robert Baron (industrialist). https://en.wikipedia.org/wiki/Robber_baron_(industrialist)

Wikipedia. (n.d.). The Gilded Age: A tale of today. https://en.wikipedia.org/wiki/The_Gilded_Age:_A_Tale_of_Today

The Revolutionary 20th Century

T he term "survival of the fittest" was originally coined in Charles Darwin's book *On the Origins of Species*, published in 1869. This simple formula could apply to the entirety of the 20th century. The numerous international conflicts reached their first peak in the "Great War" of 1914–1918, the first war fought almost exclusively with mechanical weapons. This war claimed more than 40 million lives worldwide. In the war's wake came the Spanish flu, which became the deadliest pandemic in recorded history thus far, claiming far more victims than the war itself. The death toll from the Spanish flu is estimated to be at least 50 million worldwide. After the fabulous and crazy "Roaring Twenties" came another shock — the rise of a genocidal German dictator, the Austrian-born Adolf Hitler. World War II (1939–1945) was another human catastrophe of hatred, industrial killing, and other atrocities on an unprecedented scale. As many as 85 million people lost their lives in the brutal conflict, about 3% of the world's population.

In 1916, US investors had already bet two billion dollars on the Entente's victory; that is how significant their investments were to win the war and to become the most powerful nation on the planet. In today's money, this would be equivalent to more than $500 billion. This action by the capitalist world — against the political will of US President Woodrow Wilson, who wanted to stay out of the war with his "Peace Without Victory" policy — had made the US commitment too great. During World War I, the US became the largest supplier of goods and, as a result, the largest economy in the world. After the war ended, the US entered the Depression of

1920–1921, during which total industrial production fell by about 30%. America had to worry about its own economic future, so it did not pursue President Wilson's ambition of working more closely economically and politically with France, the United Kingdom (UK), Germany, and Japan. Only the US would have been capable of becoming the anchor of a new world order. The ensuing world economic crisis — initiated by Black Thursday, October 24, 1929 — dominated the 1930s.

Figure 11. Migrant Mother (1936) by Dorothea Lange. (Public Domain)

A new world order is exactly what the dictators of the 1930s wanted to establish. Hitler and Stalin both saw themselves as radical modernizers. They modernized the world in their own way with new, radical political systems, but those systems were based on old values and old enemies. Today, Vladimir Putin, the dictator of Russia, is using similar but even more radical tactics in his pursuit of similar political goals. Putin wants to "make Russia great again." Good becomes evil, and evil becomes good. Putin's Russia had one main goal in the war against Ukraine — the destruction of the fascists in Kyiv — even as the true fascists ruled Russia.

In the case of Nazi Germany, its main goals were the destruction of the "global Jewish conspiracy" and what they called "Cultural Bolshevism." This term from the Nazi language was characterized by Carl von Ossietzky, a Nazi opponent, in 1931 as follows: "Cultural Bolshevism is what the actor Chaplin does, and when the physicist Einstein claims that the principle of the constant speed of light can be asserted only where there is no gravity, then that is also Cultural Bolshevism." In 1933, Ossietzky became one of the first inmates of a German concentration camp. In 1936, he was awarded the Nobel Peace Prize. In 1938, Ossietzky died from the brutal beatings of the Gestapo, in combination with tuberculosis that he had contracted in the concentration camps.

The Holocaust — the systematic extermination of the Jewish people — was an essential component of the new Nazi world order. From the Nazis' point of view, the Holocaust ("Shoah") was an inevitable step toward becoming and remaining the only true leading power in the world after the downfall of what they called the "Jewish-Bolshevik world order." Therefore, it was also absolutely necessary to wage a war against the US — and win it. In this respect, Adolf Hitler was probably very pleased to declare war on the US immediately after the Japanese attack on Pearl Harbor in December 1941.

After the unprecedented atrocities of World War II, both internally and externally, Germany finally capitulated. As happened after the end of World War I, the US came to the rescue. In 1918/1919, the US lent Germany large sums of money, but then withdrew. After 1945, as part of the European Recovery Plan (also known as the Marshall Plan), a lot of "real" money flowed into the devastated economies of Europe, especially Germany. Germany was divided into two states, West Germany and East Germany (the German Democratic Republic). West Germany (the Federal Republic of Germany) had to be quickly stabilized by American forces, largely to contain Soviet communism, which the US feared would push further west.

Figure 12. A photograph showing the destruction of Hiroshima (with an autograph of "The Enola Gay" Bomber pilot Paul Tibbets). (Public Domain)

According to the motto "the enemy of my enemy is my friend," West Germany was granted loans amounting to $14 billion, which is equivalent to about $150 billion today. The reconstruction of the country's almost completely destroyed economy succeeded within a decade. However, the German "economic miracle" of the 1950s came at a high price. In West Germany, the land of the perpetrators, no one wanted to question those who were now back in office. Those who had survived were all just "good Nazis," as they all claimed to have hated Hitler but were too afraid to speak out. For decades, almost no one was held responsible for the war and the implementation of Nazi ideology in everyday life — not for the euthanasia that had killed tens of thousands, not for the Nuremberg Laws that had pushed Jews out of their professions and out of society, and not for the Holocaust, the murder of millions of people of Jewish faith and origin.

After 1945, the world had to be consolidated under a new order as quickly as possible. Under the guise of large and seemingly generous financial support, the US was finally able to spread its power throughout the world. This was a double-sided embrace. On the one hand, the US military complex became the strongest in the world, which allowed the US to secure its power. On the other hand, the new US-led world order was consolidated by an ever-expanding global economy under American leadership, albeit with disastrous ecological consequences. However, ecological and environmental awareness did not matter in the former Soviet Union or its sphere of influence either.

Until recently, almost no one really seemed to care about the fatal environmental damage being done to the globe. The prerequisite for the new world order and the cementing of American supremacy was, after all, ever-increasing prosperity. Other parts of the world were struggling for power and progress. Environmental damage caused by this extractive approach was negligible or unimportant; somehow, nature would fix itself. This naive calculation did not work out for nature or for mankind. It could never have worked out when driven by a mindset of perpetual growth. Constantly increasing the gross national product does not make people happy, and it does not protect the environment without technologies geared to do so by all means. Likewise, it does not protect against mad presidents (Donald Trump) and dictators (Putin), who, in their madness, try to change the world order yet again, but in a retrograde direction, according to old models. Today's intersectional and global crisis has causes that are deeply rooted in human history. The crisis is global and omnipresent, and it can only be solved with truly new foundations and principles of human coexistence.

There were many great revolutions in the 20th century. First, there was the October Revolution (Bolshevik Revolution) in Russia in 1917. After these "Ten Days that Shook the World," the Russian Bolsheviks tried to turn the Marxist ideas of an egalitarian society into reality. This approach failed quickly and was replaced by the "dictatorship of the proletariat." In reality, this became the dictatorship of Stalin and all other Soviet rulers except Gorbachev.

The German Revolution of 1918–1919 overthrew the monarchy of Kaiser Wilhelm II. The war-loving monarchy was replaced by a politically ambitious, but unstable, Weimar Republic, which was unloved and radically opposed by the so-called "old forces." The 12 years of Nazi horror and the World War II emanating from Germany followed from 1933. The Cuban Revolution (1953–1958) was an armed uprising led by Fidel Castro against President Batista, a ruthless dictator supported by the US government. In 1956, the Hungarian revolution against the strategic domination of the USSR ended in a bloodbath inflicted by Soviet troops. Some 200,000 Hungarians were forced to flee their homeland. In 1989, the collapse of Soviet communism led to uprisings and revolutions in Poland, East Germany, Romania, Czechoslovakia, and elsewhere. The reunification of Germany was perhaps the most important geopolitical result of the final implosion of the Soviet Union.

The most important social transformations in recent history have been the sexual and digital revolutions. "Love, peace, and happiness" was the slogan of sexual emancipation in the 1960s. This meant the acceptance of sex before marriage, homosexuality, and transgender rights. The birth control pill, the legalization of abortion, and multicultural living were quickly integrated into the modern lifestyle. The digital revolution marked the transition from analog to digital technology. Computers and cell phones almost instantly became everyday commodities. They have fundamentally changed our lifestyle, our way of living together, and our perception of the world. Artificial intelligence (AI) is the next inevitable step in this progression and has the potential to change us radically, eventually leading us to lifestyles that might help us overcome our imperfections.

Figure 13. A young woman offers a flower as a symbol of peace to a military police officer at the March on the Pentagon, 1967.

Perhaps the most influential and simultaneously devastating revolution, in terms of its impact on the world's climate, has been the financial revolution. The global liberalization of markets, total privatization, and the resulting abandonment of the commons, deforestation, and global pollution are all results of the fact that the only measure of value is now capital and its accumulation. The financial crash of 2008, the recession that followed, and the complete limitlessness and "freedom" of world markets have made capitalism truly "free." You can do anything now, provided it is profitable. There are no limits or restrictions on a global scale. Everything is allowed. If we let this play on until the final act, the sheer existence of humanity and the planet itself might be able to watch the iron curtain fall in real-time.

In 1972, *The Limits to Growth: A Report for the Club of Rome's Project on the Predicament of Mankind* was published. This "little book of strong ideas" pointed to possible ecological and economic collapse within a century if "business as usual" continued with further growth and, thus, greater CO2 input into the atmosphere. A few years later, in 1977, James Black, a scientist at Exxon, told oil company executives that the burning of fossil fuels would most likely fundamentally alter the global climate. Unfortunately, this and many other warnings about the effects of fossil fuels and CO2 on the climate went unheeded. It took two more generations for climate change awareness to be raised high enough for people to take to the streets en masse. The youth movement, Fridays for Future, finally did that. They started their protest actions at a time when it had become clear to (almost) everyone that the climate was indeed changing, and that the entire planet, and humanity's survival, was in danger. There are still some people who do not understand the situation because they do not want to understand it. Hopefully, they will wake up soon.

If you look at the historical roots of the discussion on climate change, you can quickly come to the conclusion that all data and the needed reactions have been on the table for quite some time. The foreword of the introduction to *The Limits to Growth* was written by the then Secretary General of the United Nations, Hon. U Thant, as early as 1969:

> *I do not wish to seem overdramatic, but I can only conclude from the Information that is available to me as Secretary-General, that the Members of the United Nations have perhaps ten years left in which to subordinate*

their ancient quarrels and launch a global partnership to curb the arms race, to improve the human environment, to defuse the population explosion, and to supply the required momentum to development efforts. If such a global partnership is not forged within the next decade, then I very much fear that the problems I have mentioned will have reached such staggering proportions that they will be beyond our capacity to control.

The 20th century had it all. The catastrophe of the two great wars and the world economic crisis were savage blows that shattered countless people's lives. Then came the Cold War, with its nuclear threats and the global division into seemingly irreconcilable Eastern and Western blocks. China then made its great leap forward and became the world's workbench and, thus, its biggest polluter. Now, the catastrophe of a global climate crisis is clearly on the horizon. Is this just a claim by doomsday fans? Perhaps, but in a study published in the journal *Biological Reviews*, scientists gave their article this title: "The sixth mass extinction: Fact, fiction, or speculation?" The study relies on the fact that current global warming is already causing a "biodiversity crisis of increasing extinctions and declining populations," which has been met with public incomprehension because "some do not accept that this amounts to a sixth mass extinction." Scientists found that since 1500 AD, 7–13.5% of the roughly two million known species have already gone extinct.

References

Big Think. (2022). What was the most important political revolution of the 20th century? February 22. https://bigthink.com/the-past/revolution-europe-russia-china/

Black, J.F. (1978). The greenhouse effect. Exxon Research and Engineering Company, June 6. https://insideclimatenews.org/wp-content/uploads/2015/09/James-Black-1977-Presentation.pdf

Cowie, R.H., Bouchet, P., and Fontaine, B. (2022). The sixth mass extinction: Fact, fiction or speculation? *Biological Reviews*, **97**(2): 640–663. https://onlinelibrary.wiley.com/doi/epdf/10.1111/brv.12816

Darwin, C. (1859). On the origin of species. https://www.vliz.be/docs/Zeecijfers/Origin_of_Species.pdf

Digital History. (n.d.). Twentieth century revolutions. https://www.digitalhistory.uh.edu/disp_textbook.cfm?smtID=2&psid=3176

Encyclopaedia Britannica. (n.d.). Roaring twenties. https://www.britannica.com/topic/Roaring-Twenties

IMDb. (n.d.). Don't Look Up. https://www.imdb.com/title/tt11286314/

Meadows, D.H., Meadows, D.L., Randers, R., *et al.* (1972). *The Limits to Growth.* https://ia802201.us.archive.org/9/items/TheLimitsToGrowth/TheLimitsToGrowth.pdf

National Air and Space Museum. (n.d.). Boeing B-29 Superfortress "Enola Gay". https://airandspace.si.edu/collection-objects/boeing-b-29-superfortress-enola-gay/nasm_A19500100000

National Park Service. (n.d.). The atomic bombings of Hiroshima and Nagasaki. https://www.nps.gov/articles/000/the-atomic-bombings-of-hiroshima-and-nagasaki.htm

Our World in Data. (n.d.). How many people die from the flu? https://ourworldindata.org/influenza-deaths

Peterson Institute for International Economics. (2018). What is globalization? And how has the global economy shaped the United States? October 29. https://www.piie.com/microsites/globalization/what-is-globalization

The Truman Library. (n.d.). Decision to drop the atomic bomb. https://www.trumanlibrary.gov/education/presidential-inquiries/decision-drop-atomic-bomb

Wikipedia. (n.d.). Marshall Plan. https://en.wikipedia.org/wiki/Marshall_Plan

Wikipedia. (n.d.). On the Origin of Species. https://en.wikipedia.org/wiki/On_the_Origin_of_Species

Wikipedia. (n.d.). Russian Revolution. https://en.wikipedia.org/wiki/Russian_Revolution

Wikipedia. (n.d.). The Holocaust. https://en.wikipedia.org/wiki/The_Holocaust

The Ego and the Common Good

I
n recent decades, casino capitalism, human greed, and globalization
have led to an enormous overemphasis on the ego. People today tend
to trust only themselves and to concentrate their actions on the
realization of only their own goals and dreams. Progress toward any sort
of global community has suffered greatly from this development. "Be
yourself and do whatever you like. Do not take care of anything and let
others take care of themselves" — these are the new rules. However, both
possibly and realistically, there might be another way.

"Commons" are the natural and cultural resources that are accessible to
all members of a community. These include air, water, and earth. Commons
can also be defined as resources managed by communities for their own
purposes. The term commons has a long history and goes back to the
traditional English legal term for common land. This term has its roots in
the Roman *res communis*, meaning "things that can be used and enjoyed
by all members of the community." Especially in southern Europe, commons
have largely survived to the present day.

There are also cultural commons, of which the Internet platform
Wikipedia is a well-known modern example, as well as digital commons,
such as free software and open-source hardware. Public spaces in cities
become urban commons when citizens become politically active there,
for example. Community gardening, along with inner-city agriculture
on rooftops and in other cultural spaces, adds to the diversity of potential
urban commons.

Global commons have suffered severely from increasing global pollution. These are areas that are not under the direct control or jurisdiction of any state but can be used freely by all countries, companies, and individuals. These global commons include the high seas, airspace, outer space, and cyberspace.

Figure 14.   Dark clouds of factory smoke obscure Clark Avenue Bridge, NARA 550179.

If you look at historical examples, it is obvious that people both recognize and appreciate the value and advantages of collective property. Even today, the commons model could be a viable alternative. People could form new communities to jointly own and manage a resource (e.g., a piece of land). In fact, the number of such cooperatives is rapidly increasing in many countries, as being organized in a community has many advantages — it is cheaper, and you have direct access to the goods that you really want, need, and personally produce. Of course, there is always a so-called social dilemma associated with this: Who will be allowed to use a common good — when and how much of it — and will it be good for the community? Thus, you have to find the right balance within the community between your own needs and the benefit of all the people in your community. This is a major social task that needs to be solved both locally and globally.

A commons-based society is a society that values what we share as much as what we own. This means a transition from a market-based system to one that prioritizes social justice, environmental protection, and democratic citizen participation. In a society based on commons, it is still possible to apply market-based solutions, as long as they do not undermine or even destroy the operation of the commons.

A "commons society" would complement the fundamental focus on competition in our so-called liberal world with other legal principles and social structures. These would promote cooperation rather than increase competition. As the hyper-modern liberal system has led to market collapses, climate crises, and hyperactive (and sometimes frightening) states of mind, the commons model opens up a solid and proven pathway to social regeneration. Now, make no mistake... this is not just a temporary form of development. A shift to a commons-based society would mean a complete paradigm shift in the way our society functions. Put more simply — it is about treating the Earth better than we have in the past by taking responsibility for it.

References

Commons-Institut. (n.d.). Commons und Commoning. https://commons-institut.org/commons-und-commoning (in German)

Harvard Business School. (n.d.). Tragedy of the commons: What it is and 5 examples. Harvard Business School Online. https://online.hbs.edu/blog/post/tragedy-of-the-commons-impact-on-sustainability-issues

International Association for the Study of the Commons. (n.d.). About the Commons. https://iasc-commons.org/about-commons/

International Union for Conservation of Nature. (2016). New momentum for the global commons. Study, October 17. https://www.iucn.org/news/gef-iucn-partnership/201610/new-momentum-global-commons

On the Commons. https://www.onthecommons.org/about-commons/index.html

The National Trust. (n.d.). What are commons? https://www.nationaltrust.org.uk/discover/nature/what-are-commons

YouTube. (2018). What are commons? October 9. https://www.youtube.com/watch?v=WjUyfV06d7Q

World War III

I magine: It's war and you're already in the middle of it. Pop culture pastimes like computer games speculate about future nuclear catastrophes, the accompanying devastation, horrifically brutal terrorists, and other doomsday scenarios. After a devastating World War III, all major cities on Earth will be destroyed, and there will be only a few remnants of civilization — and, of course, there are always devastating gunfights with evil aliens. These doomsday scenarios obviously offer fun escapes and entertainment to millions of fans, but what if we are already in such a war-like state? There are certainly reasonable doubts about the peacefulness of our time. Imagine a military conflict between the US and China over territories in the South China Sea, including Taiwan, or perhaps an all-out war scenario between Russia and the EU, including the participation of NATO (North Atlantic Treaty Organization) troops.

On an even larger scale, imagine a war-like situation between mankind and nature. We little earthlings are ravaging the earth day after day, cutting down entire forests, polluting the oceans with plastic waste, and more. We mine, dispose, and destroy without regard for the impact. We use heaters, cars, and planes, contributing to a growing, asymmetrical battle in which one side (humans) is unaware of the potential threat from the other side (nature) and may not learn of the inevitable outcome until it is far too late.

When you find yourself in a situation too big to handle, you may think that you are safe for too long — until it is too late. Perhaps that is the case right now. The challenge is global, but the necessary ideological shift must be both global and individual. Obviously, humanity is not able to act as a whole. That is why systems like democracies actually exist, to solve these "global" problems through creative cooperation. But remember, there is no school for politicians.

Anyone who is able to speak cleverly, has the right ideas and a smiling face, or, better yet, is a member of the right party may be elected to take on the most responsible tasks of the state. There are examples of this crazy situation spinning entirely out of control. Donald Trump is one of the biggest and most famous liars in the history of the world, yet he had access to the red button to start a nuclear war. Boris Johnson is a political clown. Putin apparently has no goals other than to overrun his neighbors and trample the West. Xi Jinping wants China to be No. 1 in the world. They all got elected or came to power because many people voted for them or were persuaded to vote for them. That says something about people... they are easy to trick.

Throughout history, there have always been situations where even true democracies needed charismatic, assertive, and strong leaders. In Athens, the cradle of democracy, institutionalized tyranny was possible after everything else had failed and people were finished with all the empty words of their politicians. An example of this is the tyranny of Peisistratos in Athens. The wealthy Peisistratos accused his enemies of trying to assassinate him, and thus — along with the help of an army of foreign mercenaries — he and his sons gained power for decades. His most important achievement was the unification of Attica, which laid the foundation for Athens' later dominance. Peisistratos was a modern populist who curtailed the privileges of the aristocracy and financed bread and games for the people — and it worked.

It is no surprise that many people today sympathize with authoritarian-ruled countries like China. Totalitarian political systems have an advantage because they are able to respond in a single step to challenges that arise, whether from within the country or from abroad. Twenty-five centuries after the end of the Greek tyrannies of the Peisistratids, we, the demos ("people"), are facing a catastrophe of unimagined magnitude and unpredictable duration and consequences. Recently, we have all seen how critical our system must be to avoid tragic failure when solving huge problems such as the COVID-19 crisis.

In 1947, Winston Churchill was voted out of office as prime minister shortly after winning World War II. Of democracy, he said, "Indeed it has been said that democracy is the worst form of Government except for all those other forms that have been tried from time to time." But perhaps right now is a time like no other before. Maybe that is why we need to do things that no one has thought of doing before. Perhaps doing these unprecedented new things is the only way to save enough lives to ensure the survival of the human species. We do not know all that yet, but we must do everything we can now to prevent the worst-case scenario that is rapidly emerging.

We know from historical experience that fundamental and rapid changes in both technology and society are possible at any time. When there is a great demand for necessary changes, the pressure increases very quickly, and then suddenly, there is a lot of money for these changes. Thus, entirely new technologies and the accompanying societal changes are entirely possible within historically short periods of time. Today, in what many scientists describe as a truly critical situation, there are already many funding opportunities for further development in many areas. We should, therefore, by no means give up hope of overcoming the global climate crisis in good time.

References

Hellenica World. (n.d.). Peisistratos. https://www.hellenicaworld.com/Greece/Person/de/Peisistratos.html (in German)

International Churchill Society. (n.d.). The worst form of government. https://winstonchurchill.org/resources/quotes/the-worst-form-of-government/

Wikipedia. (n.d.). World War III. https://en.wikipedia.org/wiki/World_War_III

The World in 2040

An excellent resource and a good read to get a realistic picture of the near future is "Global Trends 2040," published by the US National Intelligence Council (NIC). NIC is an organization that bridges the United States Intelligence Community (IC, a group of US government intelligence agencies) and policymakers in the US. Its Global Trends report is published every four years and draws on intelligence from a variety of sources, including experts from academia and the private sector. The report is presented to the incoming US president as the basis for a long-term strategic policy assessment. The most recent report was published in 2021 and is entitled "A More Contested World."

The report predicts that global challenges (i.e., climate change, global pandemics and other diseases, financial crises, and technological disruptions) will become more frequent and more intense in almost all regions and countries of the world. The opportunity to address these challenges will be complicated due to increasing fragmentation within communities, states, and the international system.

According to the NIC 2021 report, transnational challenges and emergent fragmentation are outstripping the capacity of existing systems and structures, raising a new problem of imbalance. At all levels, there is a growing mismatch between the challenges and needs on the one hand and the systems and organizations that could address them on the other. There is another problem — the existing international system. All governments, organizations, and alliances seem ill-prepared to deal with these complex and growing challenges.

The entire world witnessed the COVID-19 pandemic and saw the failure of almost all health organizations and policymakers to respond quickly, reliably, and adequately to the pandemic's fatal spread. Such crises will continue in the future and will intensify as more disasters occur on a larger scale. Unfortunately, policies within states are likely to become even more volatile and contentious. No region, ideology, or government is immune to human incapacity, and no one has the answers to all the issues we now face.

Fortunately, there is light at the end of the tunnel. The coming responses to the climate crisis will be a necessity and a challenge for all actors, but we can all profit from the process. Climate change will require all nations and societies to adapt to a warmer planet. Some actions are as inexpensive and easy to implement as restoring mangrove forests and peatlands or increasing rainwater storage. Others are more complex, such as transforming the heat and transportation sectors, building massive marine conservation facilities, or relocating large populations. All in all, the picture we face is radically different from that of a glorious, prosperous future. By 2040, the international system will likely be without direction, chaotic, and volatile as international rules and institutions will be largely ignored by major powers such as China, along with regional actors and non-state actors (National Intelligence Council, 2021, p. 5).

To cite a few figures: In 2020, about 270 million people lived in countries to which they immigrated. In the future, many countries, especially in the so-called Global South, will see increasing numbers of people fleeing conflicts, environmental disasters, and economic decline. Most people will want to immigrate to richer, safer countries in Europe, Asia, and North America. It is quite possible that the number of migrants will at least double in the next two decades. The disasters that could occur along so many millions of paths are being foreshadowed by what is happening in the Mediterranean today.

The NIC 2040 report provides a more or less detailed look at many areas of future politics and society, including war scenarios with the (limited) use of nuclear weapons due to a competitive geopolitical environment. As you read, remember that this report was not written by crackpots or conspiracy theorists but by scientists reporting to the incoming president of the United States.

References

National Intelligence Council. (2021). Global trends 2040: A more contested world. Report, March. https://www.dni.gov/files/ODNI/documents/assessments/GlobalTrends_2040.pdf

Wikipedia. (n.d.). United States Intelligence Community. https://en.wikipedia.org/wiki/United_States_Intelligence_Community

PARADISE Lost —
Burning Hope

T he Minoan civilization perished in an inferno, the likes of which no advanced civilization had ever experienced before. The explosion of an entire island — then known as Thera, the remains of which are now called Santorini — and the ensuing tsunami engulfed all the major coastal sites of that maritime people. The effects of the eruption, which probably occurred around 1430 BC, so weakened the most advanced civilization of the time that the Minoans were quickly conquered and subjugated by the Mycenaeans, a highly militarized culture from mainland Greece.

Pompeii and Herculaneum, two thriving cities near present-day Naples, were buried under a five-meter-high carpet of volcanic ash when Mount Vesuvius erupted in 79 AD. Almost the entire population perished. It is estimated that 15,000 to 20,000 people died across the entire area, which was hit by massive pyroclastic waves and ash rain.

In the course of time, i.e., during the written history of mankind, there have been many similar disasters, almost all of them due to natural causes. In recent history, however, there have been more and more man-made disasters, such as Chernobyl, Harrisburg, Fukushima, or the oil spill after the explosion of the Deepwater Horizon oil rig. In the last decade, more and more severe hurricanes, floods, and/or droughts have taken many lives and incurred immense recovery costs.

In recent years, two towns with iconic names have been hit by wildfires — the first was Paradise, near Santa Barbara in California (US), and then

Hope in British Columbia (Canada). The small town of Hope burned first and was then hit by floods that almost completely submerged the town.

Figure 15. "Camp Fire," November 8, 2018 (near Paradise). (Source: NASA, Joshua Stevens)

The more frequently such catastrophes occur, the more people begin to actively worry about climate change. It seems that nature is sending out wake-up calls for humanity to finally pay attention. It seems to be working, even if the price is painfully high. In July 2021, the Ahr Valley in Germany was hit by a cataclysmic flood. It killed 117 people and devastated the entire region for years to come. The cost of this single disaster is estimated to be well over 30 billion euros.

People are beginning to realize how costly climate catastrophe is, both in terms of human lives and money. We must understand and accept that we have to live with nature rather than seek to control it.

As a species, we have already solved a large number of massive problems, but unprecedented challenges lie ahead. We have survived world wars, deadly pandemics, and many other disasters, but now there is a new enemy involved. We must deal with our egos and protect ourselves from ourselves — from our self-defeating urge to consume more and more, to expand, and to dominate the whole world.

There are moments in the history of mankind that are crucial for the further development of civilization. This is one such turning point. History has taught us that our capabilities enable us to overcome even the most difficult obstacles. The greatest challenge of our time is the climate crisis. It is the ultimate threat to the entire ecosystem of our planet, and it is time to fight for our survival. We must cultivate new ideas for societal function and develop those ideas for practical use. The best thing about the many challenges of global climate change is that all the necessary solutions already exist. All we have to do is to join forces, use our intelligence, and direct our resources to implement these solutions, designs, prototypes, and projects.

We need to change right now and continue changing as we move into the future, learning every step of the way. In the process, we need persistent optimism, endless courage, and a constant willingness to adapt to dynamic challenges. On the road to a sustainable world, we have to fight for our survival on a healthy planet. First, we need to win the race to a world with a balanced CO_2 budget ("Race to Net Zero") by 2050.

The cultural background into which our children and grandchildren are born includes various forms of governance and legislation, along with all the associated structures, institutions and administrations, rules, resulting habits, perceptions, and interpretations. We ourselves are part of this culture, which we have internalized and almost unconsciously passed on.

However, something deeper has disintegrated in the past two or three generations, and we need to change that quickly.

References

European Forest Fire Information System. https://effis.jrc.ec.europa.eu/

Labaudiniere, M.S. (2012). Three Mile Island, Chernobyl, and the Fukushima Daiichi Nuclear Crises: An argument for normal accident theory. Thesis, Boston College. https://dlib.bc.edu/islandora/object/bc-ir:102349/datastream/PDF/view

The Guardian. (n.d.). Wildfires. https://www.theguardian.com/world/wildfires

The Metropolitan Museum of Art. (n.d.). Mycenaean civilization. https://www.metmuseum.org/toah/hd/myce/hd_myce.htm

Thera Foundation. (n.d.). Is Santorini the lost city of Atlantis? https://therafoundation.org/santorini-volcano/atlantis-santorini-theory

United Nations Climate Change. (n.d.). Race to zero campaign. https://unfccc.int/climate-action/race-to-zero-campaign

Wikipedia. (n.d.). Nuclear and radiation accidents and incidents. https://en.wikipedia.org/wiki/Nuclear_and_radiation_accidents_and_incidents

Wikipedia. (n.d.). Pompeii. https://en.wikipedia.org/wiki/Pompeii

World History Encyclopedia. (n.d.). Mycenaean civilization. https://www.worldhistory.org/Mycenaean_Civilization/

II. Players and Models

Crisis? What Crisis?

Yet we cannot avoid noting with concern how today — and not only in Europe — we are witnessing a retreat from democracy.

— Pope Francis in 2021

The international order is in crisis. The most obvious signs of this are the near collapse of the global financial system in 2008, the rise of right-wing movements in Europe and the United States (US), the Brexit vote, and the election of Donald Trump as the US president in 2016. In addition, Russia appears to be ruled by a dictator with a penchant for waging wars and threatening to use nuclear weapons. These events challenge the international order of global governance, economic freedom, multilateral trade, and security cooperation. It could take decades to reestablish that balance.

Today's financial crisis (despite booming share values) has its roots in the operating principles of the postwar world economy and financial sector under American leadership. In the years leading up to the financial crisis of 2007/2008, financial markets were deregulated, and the real estate market boomed in the US. When the financial bubble burst in 2008, the effects were devastating. After a $700-billion bank bailout in the US alone, the global financial system gradually stabilized, but many Americans had lost their homes, and the unemployment rate in the US rose to a staggering 10%.

After 2012, the financial crisis evolved into a political crisis of the entire post-war global system. The shaky state of Greece nearly failed, and other European countries (Ireland, Portugal, and Cyprus) lost a decade of prosperity. In the US, post-bailout financial stagnation eventually led to Donald Trump's election victory. In his populist-alarmist jargon, Trump pointed the finger at a global structure that had robbed the working class and put money in the pockets of a handful of large corporations and political entities. In this respect, Donald Trump's presidency could rightly be described as collateral damage of the financial crisis. Currently, Trump is trying to move back into the White House in 2024 with the help of the Republicans, although he is facing strong legal headwinds.

In Russia, the situation is even more abysmal. There, a bizarre autocrat heads a pseudo-state ruled by himself and his oligarchic puppets, where he and those same puppets act as modern-day robber barons, plundering the Russian state's funds and properties. Wars are instigated as a diversion and/or last resort. In this absurd scenario, Putin portrays himself as Russia's savior, forced to fend off the onslaught of fascist drug addicts from Ukraine (and possibly other regions) in the name of the Russian people and freedom. Lies rule in Putin's empire, just as Trump's imaginary empire was built on deception.

The term "woke" shows how even a single word can be misused by political and social groups in this ongoing political chaos. Originally, "woke" referred to an "awareness of a lack of social justice and racism." It developed in the 1930s in the US and has been increasingly used to refer to a "politically awake" consciousness, especially against discrimination, since the early days of the Black Lives Matter movement. However, the term is now widely being used by conservatives as a synonym for what they see as overly progressive political activism that threatens to destroy the values, ideals, and freedoms on which they believe America was founded. This example illustrates how deep the divide in (and not just the American) society has become, simply in the use of individual words and terms. The conservative right is driven by crafting narratives that it can abuse for its own purposes. So, both sides of the political spectrum need to stay awake!

Fortunately, every crisis implies future reconstruction. After some of Trump's supporters attempted a *coup d'état* by attacking the US Congress, the juridical system took action. That mad political adventure ended with the inauguration of Democrat Joe Biden, though the danger has not passed forever. President Biden is trying to steer America back into friendlier waters in the (Western) world and trying his best to "rebuild better." We have yet to see whether New Deal-style legislation will really become the centerpiece of Joe Biden's domestic policy in the US.

Biden's draft program "Build Back Better" was subsequently turned into the "Inflation Reduction Act of 2022" over the course of the negotiations. This reduced the original draft, especially in the area of climate protection, to such an extent that the threat of further global warming would become even more realistic. In a globalized world, the US decision would mean that other countries are likely to use the same arguments of conservative opponents in the US Senate. That being said, China, the world's biggest polluter as a nation, has already announced its intention to become climate-neutral (net zero) by 2060. This is a definite sign that China is taking serious measures against climate change that other nations could follow.

There is another, much greater threat to the reasonable development of climate change policies and measures: the growing number of people around the world who have lost their belief in governments and will oppose any actions from authorities. During the COVID-19 pandemic, anti-vaccination activists threatened doctors and hospital staff and even plotted to kill people they blamed for the so-called "vaccine dictatorship." Their protests were accompanied by calls for "freedom" and "resistance." By now, these groups are well organized internationally, and when bolstered by key figures like Andrew Wakefield, a trained physician, supporters feel that they are on the right side of the argument. It is not much different for climate change deniers. From the denial camp, we still hear that there is no consensus among scientists about climate change, that humans can easily adapt to freak weather events, that this sort of thing has happened before, or that it is too late to do anything about it anyway. This stubbornness will likely continue for a long time, and it may become increasingly difficult to engage with those who believe in this nonsense.

However, if you can read, the crisis seems so simple to accept. As summarized in the IPCC 2022 Final Report, "Human-induced climate change, including more frequent and intense extreme events, has caused widespread adverse impacts and related losses and damages to nature and people, beyond natural climate variability. Some development and adaptation efforts have reduced vulnerability. Across sectors and regions, the most vulnerable people and systems are observed to be disproportionately affected. The rise in weather and climate extremes has led to some irreversible impacts as natural and human systems are pushed beyond their ability to adapt." (IPCC, 2022, Headline statements, B.1).

Reference

Intergovernmental Panel on Climate Change. (2022). Climate change 2022: Impacts, adaptation and vulnerability. Working Group II contribution to the sixth assessment report of the Intergovernmental Panel on Climate Change, Pörtner, H.-O., Roberts, D.C., Tignor, M.M.B., et. al. (eds.). https://www.ipcc.ch/report/ar6/wg2/

The Green Revolution

The data is clear. The moment is now.

— John Doerr, *Speed and Scale* (2021)

Since the Industrial Revolution, man seems to have been intent on ruining the environment. In the past, the Romans cut down the trees across half of Italy to build their fleets, roads, and houses. Creating farmland by burning forests is still common around the world, for example, in the Amazon rainforest. Atomic bombs were tested at Bikini, Mururoa, and in the Australian desert. The fallout from the bombs contaminated all life in the area for centuries to come. There is no documented data on the greenhouse gas emissions of the two world wars, but they are certainly gigantic. According to research by Scientists4future, the Second Iraq War caused 141 million tons of CO_2 emissions from March to May 2003 — about as much as New Zealand or Cuba emits per year. When you look at the IPCC reports and the numbers, you quickly get the impression that we have already reached the end of the line. That is where we are: The "Uninhabitable Earth" is the elephant in our living room that we simply do not want to see. In fact, most people do not even want to think about it, preferring to deal with other everyday problems that seem far more manageable and immediate.

Currently, the CO_2 content in the air is 410 ppm (parts per million) compared to 290 ppm, which scientists believe is the basis for the comfortable average temperature of 15°C worldwide that people have been accustomed to for centuries. Since 1950, the annual growth rate of global

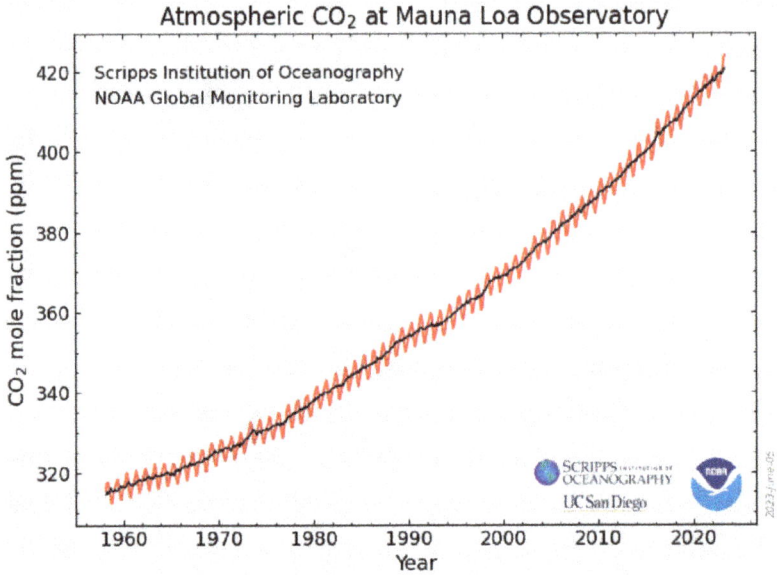

Figure 16.   Atmospheric CO2 concentration. (Source: NOAA Global Monitoring Laboratory)

carbon dioxide emissions has quadrupled, the world's population has tripled, and global energy demand has increased fivefold over the same period. All of this has resulted in enormous environmental stress that will be felt for centuries to come.

John Doerr's meticulously researched and undeniably good book *Speed and Scale* provides further figures: Humanity releases about 59 gigatons (billions of tons) of CO2 into the atmosphere each year. Each person's CO2 footprint per year must be reduced by two-thirds within the next 20 years. Don't fly. Climate Action Tracker is an independent scientific analysis platform. It measures climate action by governments against the globally agreed-upon goal of keeping warming well below 2°C and continuing efforts to limit global warming to 1.5°C. It is a collaboration of the following organizations: Climate Analytics is a non-profit organization based in Berlin, Germany, with offices in Lomé, Togo, and New York, US. It brings together interdisciplinary expertise on the scientific and policy aspects of climate change. Its activities include: synthesizing and advancing scientific knowledge on climate change science, policy, and impacts; providing scientific and policy support to the least developed countries and small island states in international climate negotiations; and reviewing and analyzing the effectiveness of national climate policies worldwide. The New Climate Institute is a non-profit institute founded in 2014. It supports research and the implementation of climate change policies around the globe, covering international climate negotiations, climate action tracking, climate and development, climate finance, and carbon market mechanisms. The institute aims to connect current research with real-world decision-making processes.

Small and modest steps will not be enough. We need well-conceived, bold, and powerful actions now, considering the magnitude of the tasks ahead. We need nothing less than a revolution — the Green Revolution. This revolution needs people with power who will enable or initiate actions to promote change and who are willing to help transform almost everything in modern society in a holistic way.

The goal of the first stage in the Green Revolution is to meet the climate targets of the Paris Agreement. In this 2015 agreement, all signatory countries pledged to limit global warming to well below 2°C, as compared to the pre-industrial era around 1880. This agreement is currently considered the "gold standard" of global efforts to limit global warming. Changes in the Earth's climate that have occurred over the past two decades demonstrate that it is imperative that the goals of the Paris Agreement be met in order to prevent the following:

- More frequent devastating weather events (heavy rains, hurricanes, etc.)
- Severe droughts leading to crop failures and global food insecurity

- Collapse of the world trading system, the increasingly vulnerable backbone of the global economy
- Deterioration of people's living conditions, especially in the poorer regions of the world
- Triggering of global migration movements of unprecedented proportions

The global financial world already understands that it will be absolutely necessary to change the goals and intentions of their companies in order to survive this global crisis. In nearly every single business model, the machines being used, the respective energy supply, and perhaps even the respective products themselves will have to be changed as a result of this crisis. This shift will continuously require a lot of money, and an all-out effort will be necessary.

Paris demands we do things differently. The time is long passed for "greening" the current system!

— Kevin Anderson, former director of the Tyndall Center for
Climate Change Research

References

Climate Action Tracker. https://climateactiontracker.org/

Climate Analytics. https://climateanalytics.org/

Cottrell, L. and Dupuy, K. (2021). We must not ignore explosive weapons' environmental impact. Conflict and Environment Observatory. https://ceobs.org/we-must-not-ignore-explosive-weapons-environmental-impact/

European Central Bank. (2021). Climate-related risk and financial stability. July. https://www.ecb.europa.eu/pub/pdf/other/ecb.climateriskfinancialstability202107~87822fae81.en.pdf

globalCarbon Atlas. https://globalcarbonatlas.org/

Grippa, P., Schmittmann, J., and Suntheim, F. (2019). Climate change and financial risk. International Monetary Fund, December. https://www.imf.org/en/Publications/fandd/issues/2019/12/climate-change-central-banks-and-financial-risk-grippa

Giuzio, M., Krusec, D., Levels, A., *et al.* (2019). Climate change and financial stability. European Central Bank. https://www.ecb.europa.eu/pub/financial-stability/fsr/special/html/ecb.fsrart201905_1~47cf778cc1.en.html

New Climate Institute. https://newclimate.org/

United Nations. (n.d.). Finance & justice. https://www.un.org/en/climatechange/raising-ambition/climate-finance

Wikipedia. (n.d.). Amazon rainforest. https://en.wikipedia.org/wiki/Amazon_rainforest

Wikipedia. (n.d.). Keeling Curve. https://en.wikipedia.org/wiki/Keeling_Curve

The Climate and the Law

Many of the laws in our world serve property — they are based on ownership. But imagine a law that has a higher moral authority… a law that puts people and planet first.

— Polly Higgins (1968–2019), "The Earth's Lawyer"

Philippe Sands (b. 1960) is a specialist in international law and director of the Centre for International Courts and Tribunals at University College London (UCL). He has written numerous books and is a regular contributor to the *Financial Times* and *The Guardian*, two British newspapers. Sands, along with other colleagues, has already drawn up plans for the enforceable legal crime of ecocide, i.e., the progressive destruction of the world's ecosystems. Similar to existing international crimes, such as crimes against humanity — established in the Charter of the International Military Tribunal (IMT) at Nuremberg in 1945 — Sands' initiative has already gained the support of European countries (France), as well as several island nations facing rising sea levels. The legal body is currently working on the definition of ecocide under international criminal law and outlining how to achieve accountability. Commenting on these activities, Sands stated: "My hope is that this group will be able to … forge a definition that is practical, effective and sustainable, and that might attract support to allow an amendment to the ICC [International Criminal Court] statute to be made."

Although the biggest polluters, such as China, the US, and Russia, are not members of the ICC, this is a start toward holding individuals, corporations, and perhaps even elected officials accountable for environmental crimes.

The panel has already proposed a definition to be included in the legal framework of the Rome Statute of the ICC: "For the purpose of this Statute, 'ecocide' means unlawful or wanton acts committed with knowledge that there is a substantial likelihood of severe and either widespread or long-term damage to the environment being caused by those acts."

One fact appears to be clear: So long as there is no suitably high carbon tax (CO2), there will be far too many fossil fuels. These resources have allowed our economy to grow, and they still do, unchecked because the cost of fossil fuels (e.g., gasoline) does not account for the damage done to nature and the environment.

A 2019 International Monetary Fund (IMF) report states that carbon taxes and similar arrangements to raise the carbon price are the most powerful and efficient tools for reducing domestic CO2 emissions from fossil fuels. A (high) tax on each ton of greenhouse gas (GHG) emissions would be the most straightforward and effective tool to incentivize energy producers to "go green." This type of "tax" (carbon pricing) would significantly reduce carbon emissions in the long run because continued use would be too costly, and, at the same time, this tax would generate significant revenues to fund environmental and eco-social activities. This method is already working well from a practical point of view. Carbon pricing in Sweden has led to a reduction of emissions by more than 25% over the past 30 years.

References

Centre for International Courts and Tribunals. https://www.ucl.ac.uk/international-courts/centre-international-courts-and-tribunals

E3G. https://www.e3g.org/

International Criminal Court. (1998). Rome statute of the International Criminal Court. https://www.icc-cpi.int/sites/default/files/RS-Eng.pdf

Parry, I. (2021). Five things to know about carbon pricing. International Monetary Fund, September. https://www.imf.org/en/Publications/fandd/issues/2021/09/five-things-to-know-about-carbon-pricing-parry

Stop Ecocide International. https://www.stopecocide.earth/

Wikipedia. (n.d.). International law. https://en.wikipedia.org/wiki/International_law

Money, People, and Institutions

ig Money has realized that keeping customers alive and working is essential to the survival of the global economy. Mark Carney is a key figure within the global financial system for the transition from fossil fuels to a green economy that Big Money is striving for in the long run, whereas conservative circles of the global money club still invest in fossil fuels and technologies. Smart and agile, Carney is an experienced banker who served as Governor of the Bank of England from 2013 to 2020. He is currently Vice Chairman of Brookfield Asset Management, one of the world's largest real estate investors, where he serves as Head of Transition Investing. He is also a member of the Board of Trustees of the World Economic Forum in Davos. Carney, born in 1965, is also a UN Special Envoy for Climate Policy and Finance. He was financial advisor to Prime Minister Boris Johnson for COP26, the Glasgow 2021 climate conference, and is also chair of the new Glasgow Financial Alliance for Net Zero (GFANZ). The common goal of the companies in the GFANZ alliance is to steer the global economy toward net-zero emissions.

Mark Carney confirmed at the COP26 conference in Glasgow that the GFANZ now includes more than 450 companies with US$130 trillion in assets. This is a truly staggering amount of total investment power in the green economy. But what does US$130 trillion (US$130,000,000,000,000) mean? It is the promise of top investors for how much they are willing to invest against the climate crisis. Whether it will really happen remains uncertain. It is currently just a huge number on paper, and one can only hope that this amount will be enough for the complete transition of all industries worldwide. At least it is a good start.

Skeptics contend that the underlying terms of the GFANZ commitments are questionable from the outset, as signatories to GFANZ are in no way obliged to stop funding the fossil fuel industry. Antonio Guterres, Secretary-General of the United Nations, stated, "There is a deficit of credibility and a surplus of confusion over emissions reductions and net-zero targets, with different meanings and different metrics." Who will verify what GFANZ investments will be used for? Who will set the standards for evaluating and analyzing Net Zero commitments?

Mark Carney loves to talk about "carbonomics" — tax expenditures to retrofit people's homes or support certain types of renewable energy. He points to financial initiatives in the European market, as well as German and French initiatives on hydrogen. The goal of the German hydrogen initiative, for example, is to reduce the cost of producing green hydrogen to the point where Germany can become a leader in the coming global hydrogen economy. In Namibia, a former colony of Germany in Southwest Africa, a huge wind and solar park is to be built. The company is sponsored by the Namibian project development company Hyphen Hydrogen Energy, as well as by the energy company ENERTRAG, based in the German state of Brandenburg. The hydrogen produced in the Namibian Tsau-Khaeb National Park is to be exported to Germany and worldwide after completion of the production facilities, which is scheduled for 2028.

A second issue is what is known as "framing." This refers to regulatory initiatives within the "real economy." These initiatives can be even more important than planned investments in terms of guidance regarding their concrete impact, both economically and socially. For example, it would be good to know when the internal combustion engine will become extinct in automotive production. Currently, this is planned for 2035, but it might happen a few years earlier. Private sector investments can be guided by such important dates, i.e., such a date gives investors the necessary clues as to where and when they should invest. Any financial decision for or against investing in a company or product should also be thoroughly assessed in terms of its impact on the climate. Only if such an assessment is positive should investments be made.

The financial agenda of the future must be shaped according to those assessments. For this to happen, the financial system needs all the information it can get about markets and market infrastructure to translate those into investment plans and, finally, action. Building on the foundations of reporting and risk management, the financial system can positively look forward and address climate change through commitments, alignment, and the necessary investments. Mark Carney stated in an IMF release, "Alignment means defining best practice net zero plans for companies and financial institutions, leveraging the valuable work already begun. Alignment also means robust assessments of the portfolios of financial institutions relative to net zero pathways."

During the United Nations (UN) summits, many world leaders come together, and at these meetings, the three issues mentioned above are always on top of the many topics to be discussed. This gives hope because these talks are usually about other, more political matters. It was like during the financial crisis, when the world's most powerful were much more willing to try more radical things simply because it was necessary. Climate policy is very much at the top of the UN's agenda now.

The principles of responsible investing are referred to as ESG (Environmental, Social, and Governance). Most asset managers believe that an economically efficient and sustainable global financial system is a necessity for long-term value creation. Such a system will reward responsible investment while benefiting the environment and society as a whole.

The Science-Based Targets initiative (SBTi) is developing the world's first standard for science-based net-zero targets in the financial sector. The initiative has already published a report to build consensus on the term "net zero." Perhaps this initiative will lead to a new set of rules for companies and institutions in the financial industry that can serve as a guide for them to keep their promises. The creation of new global standards for climate reporting is essential to achieve the net zero goals as quickly and safely as possible.

In the future, the global market itself will be extremely volatile, i.e., unstable. For investors, financial risks arise not only from the effects of global warming on the ecosystem and the planet (e.g., extreme weather) but also from so-called transition risks. These are risks that will arise from essential regulation, technological innovations, changing market dynamics, and shifts in people's consumption habits.

The actions that companies must take to limit climate change will have a significant impact on their profitability in the coming years. Investments in transformative technologies that directly contribute to the goal of GHG neutrality must be significantly increased. This includes investments in research and development and in GHG-neutral technologies that have already reached market level. TWH Zurich spin-offs Synhelion (Sunlight to Fuel) and Climeworks (Direct Air Capture) are prime examples; both offer market-ready technologies for a transition to a net-zero future. Both of these fast-growing companies now need to reach a global scale to achieve tangible success in terms of their impact on climate change.

References

Dolan, B. (2023). Namibia and Hyphen HydrogFraming effect: What it is and examples. *Investopedia*, May 11. https://www.investopedia.com/framing-effect-7371439

Gladman, K. (2022). Climate risk: An investor resource guide. Principles for Responsible Investment, January 22. https://www.unpri.org/download?ac=15605

Glasgow Financial Alliance for Net Zero. https://www.gfanzero.com/

Goldman Sach. (n.d.). Carbonomics. https://www.goldmansachs.com/intelligence/topics/index.html

Hamilton Lane. (2022). Environmental, social, and governance investment policy statement. March. https://www.hamiltonlane.com/getmedia/6f3460e4-59ac-45b9-9e0c-3e99cfa43e56/esg_policy_statement_1-2021.pdf

Hydrogen Central. (2023). Namibia and Hyphen Hydrogen Energy sign US$10 billion green hydrogen project agreement at official ceremony. June 23. https://hydrogen-central.com/namibia-hyphen-hydrogen-energy-sign-us10-billion-green-hydrogen-project-agreement-at-official-ceremony/

Hyphen Hydrogen Energy (Pty) Ltd. https://hyphenafrica.com/

Naschert, C. (2021). "Deficit of credibility": UN deepens benchmarking of corporate net-zero pledges. S&P Global Market Intelligence, November 2. https://www.spglobal.com/marketintelligence/en/news-insights/latest-news-headlines/deficit-of-credibility-un-deepens-benchmarking-of-corporate-net-zero-pledges-67395527

Office of Science. (n.d.). DOE explains... solar fuels. https://www.energy.gov/science/doe-explainssolar-fuels

Reuters. (2023). Hyphen and Namibia agree next phase of $10 billion green hydrogen project. May 25. https://www.reuters.com/business/energy/hyphen-namibia-agree-next-phase-10-bln-green-hydrogen-project-2023-05-24/

Science Based Targets. (2022). Financial sector science-based targets guidance, version 1.1. August. https://sciencebasedtargets.org/resources/files/Financial-Sector-Science-Based-Targets-Guidance.pdf

Task Force on Climate-related Financial Disclosures. (2017). Recommendations of the Task Force on Climate-related Financial Disclosures. Final report, June 15. https://assets.bbhub.io/company/sites/60/2020/10/FINAL-2017-TCFD-Report-11052018.pdf

Larry Fink and the Green Transition

arry Fink (b. 1952) is CEO and chairman of BlackRock, the world's largest asset management company with over $9 trillion in assets. In his 2021 annual letter, Fink wrote: "I believe that the pandemic has presented such an existential crisis — such a stark reminder of our fragility — that it has driven us to confront the global threat of climate change more forcefully and to consider how, like the pandemic, it will alter our lives. It has reminded us how the biggest crises, whether medical or environmental, demand a global and ambitious response."

Since 2012, Fink has sent his annual letters to the CEOs of companies in which BlackRock invests on behalf of its clients. These are considered an insight into how the financial industry will develop and what the core shifts and values in the future will be. Right now, the question is: How will market leaders define their role in society during the revolutionary shift to Net Zero? The 2020 COVID pandemic has already done its part to challenge the way businesses think. Now, larger issues are at play — from public health to climate change and social justice. They will all affect the way business will be done in the future.

In his 2021 letter, Fink laid out a blueprint for how BlackRock will deal with the "green" transition in the financial industry. He pointed out that the transition will happen faster than expected, as the impact of climate change will be greater and happen sooner than previously thought. Fink specifically suggests that a net-zero corporate strategy will be a future imperative for stakeholders and investors to ensure that a company is well-equipped to navigate the challenging transition while remaining profitable.

In addition, corporate HR (human resources) strategies should include plans to improve in the areas of diversity, equality, and inclusion.

The global shift to divest from fossil fuels is slowly gaining traction in the financial world. It began after an article published by the London-based Global Tracker Initiative revealed that major fossil fuel companies (Exxon, BP, etc.) have five times more carbon in their reserves than scientists believe the world should burn if we are to meet reasonable temperature targets. These numbers were unambiguous, so the move away from fossil fuels became more than just a trend. Of course, the fossil fuel industry fought back. BlackRock then conducted a study to get hard numbers on the risks of divesting from fossil fuels. In the conclusions of the BlackRock report were the magic words that investors love to hear: Portfolios that have divested from fossil fuels have outperformed their benchmarks (comparable companies).

Fink's 2021 letter reveals that climate change is the most important issue of our time, not only for sustainability activists but also for global trade and businesses. Fink has good reason to go green: "The climate transition presents a historic investment opportunity." More and more shareholders want the companies they invest in to reflect their positive view of ESG issues. The ambitious road ahead requires collective responsibility for sustainability and a holistic, collaborative approach. As Fink says, "Companies that do not earn this trust will find it harder and harder to attract customers and talent, especially as young people increasingly expect companies to reflect their values." The letter's key phrases clarify its objective: transition, net zero, climate, global.

Tariq Fancy is the former chief investment officer (CIO) for sustainable investing at BlackRock. Unlike Larry Fink, he believes that sustainable investing and the ESG movement will not solve global problems like climate change. In Fink's initiative, he sees a contradiction to the previously prevailing Milton Friedman orthodoxy, namely, that an investment company should focus on making profits because that is its duty when managing other people's money. Also, Fancy asks why Fink's shift to a new credo should be taken seriously — what it would really mean (and cost) if companies should have a "social purpose beyond financial performance."

If the goal of asset management firms is to maximize profits and increase shareholder value, and they all say that they are now going green, one question remains: Who will control those claims? ESG is just a word, but a clear definition is still lacking. Some so-called ESG investment funds are simply mislabeled ("greenwashing"). What is truly missing is a global discussion by economists, asset management firms, and governments to define a responsible ESG business in terms of the global green agenda.

The COVID-19 pandemic has taught us that precise, regulative government action is absolutely necessary in a global crisis. If it was necessary during the COVID-19 pandemic, then it will be even more necessary in the much larger global climate crisis. Even so, Fancy is pretty sure that Fink's words are just covert lobbying against climate legislation, saying, "No thanks, we don't need government action to fight climate change. I prefer that capitalists regulate themselves."

Fink certainly does not represent the position of the younger people who make up the Fridays for Future movement. They recognize that climate change is a historic challenge, not a short-term investment problem to work around. The fight against climate change is just beginning, and the younger generation, with their compelling ideas and beliefs, must hold fast to their inner truth.

Now, there is at least one thing that Fink's 2021 letter achieved: He brought the issue of green investment into the boardrooms of international finance. So perhaps his 2021 letter really was a significant leap forward. After all, the world's largest investment house sent a clear message to the financial world as to how they can protect themselves, the global economy, and the planet: invest in ESG companies! Hopefully, BlackRock's insights will take hold in markets around the globe, guiding more and more investors into green territory.

References

Fink, L. (2021). Larry Fink's 2021 letter to CEOs. BlackRock Inc. https://www.blackrock.com/us/individual/2021-larry-fink-ceo-letter

Fink, L. (2022). The power of capitalism. BlackRock Inc. https://www.blackrock.com/corporate/investor-relations/larry-fink-ceo-letter

GlobeScan. (2021). GlobeScan Analysis 2021: Larry Fink's 9th annual letter to CEOs and trends since 2012. https://globescan.com/wp-content/uploads/2021/02/GlobeScan_Analysis_Larry_Fink_Annual_Letters_to_CEOs_Feb2021.pdf

Huber, B.M. and Simpkins, P.H. (2021). BlackRock's 2021 CEO Letter. Harvard Law School Forum on Coporate Governance, February 14. https://corpgov.law.harvard.edu/2021/02/14/blackrocks-2021-ceo-letter/

McKibben, B. (2021). The powerful new financial argument for fossil-fuel divestment. *The New Yorker*, April 3. https://www.newyorker.com/news/daily-comment/the-powerful-new-financial-argument-for-fossil-fuel-divestment

Sorkin, A.S. and de la Merced, M.J. (2022). It's not "woke" for businesses to think beyond profit, BlackRock chief says. *The New York Times*, January 17. https://www.nytimes.com/2022/01/17/business/dealbook/larry-fink-blackrock-letter.html

Wooldridge, A. (2022). Business doesn't need a "social purpose" revolution. Bloomberg, January 18. https://www.bloomberg.com/opinion/articles/2022-01-18/larry-fink-is-wrong-business-doesn-t-need-a-social-purpose#xj4y7vzkg

Women at the Top

The financial world is still dominated by white men in dark blue suits, but in recent decades, some women have succeeded in reaching the top of the financial world.

Kristalina Georgieva (b. 1953) is currently Managing Director of the International Monetary Fund (IMF), a post she took over from Christine Lagarde. Georgieva comes from a Bulgarian family of bureaucrats and has robust international experience and knowledge of both finance and politics. At the IMF, she is in charge of overseeing the Fund's work, i.e., helping countries achieve macroeconomic stability and reduce poverty. In the future, the IMF will have to redefine its role, as the BRIC (Brazil, Russia, India, and China) countries and other developing nations are most affected by the climate crisis, to which they have contributed to a much lesser extent.

Christine Lagarde (b. 1956), a lawyer, was elected as the first woman to head the European Central Bank (ECB). The ECB is one of the most important central banks in the world. These banks are lenders of last resort, meaning that they are responsible for providing money to a country's economy when commercial banks run out of funds. Another role they play is to ensure price stability by monitoring inflation. In the US, the lender of last resort is the Federal Reserve Bank.

Christine Lagarde commented that after the 2007/2008 crisis, the IMF endorsed immediate fiscal stimulus for all major economies and a massive expansion of monetary policy. When asked about the causes of the financial crisis, Lagarde referred, not without a sense of humor, to the widespread "group thinking" in a male-dominated industry. As a consequence, she repeatedly called for comprehensive gender reform, i.e., more women in executive board functions.

As Director of the European Central Bank (ECB), Lagarde is responsible for the ongoing analysis of the potential impact of climate change. In doing this, the ECB performs so-called stress tests for some four million companies worldwide and for almost all European banks. Lagarde advocates for a green transition of the economy, carbon pricing that reflects social costs, full disclosure of climate change impacts in the banking sector, and the completion of the EU Capital Markets Union. In addition, Lagarde has frequently emphasized the most critical issue in corporate ESG assessment — consistent and verifiable data disclosure. As a result, the ECB has already accepted a large number of sustainability-related bonds as part of its asset purchase program.

Janet Louise Yellen (b. 1946) is an American economist, educator, and government official. She has been Secretary of the Treasury of the United States since 2021. Previously, she served as the 15th Chair of the Federal Reserve. She is considered a Keynesian economist who advocated stimulus programs in the years following the 2007/2008 crisis, and she believes that "government has a duty to reduce poverty and improve lives." In a report released in October 2021, Yellen called the financial risks associated with climate change "a unique, existential risk for the planet that will affect every aspect of our lives and the lives of our children." She advocates for immediate economic adjustments to climate change in order to enable a transition to a net-zero economy.

The financial crisis of 2007/2008 hit the US market hard. As a result, the Obama Administration established the Financial Stability Oversight Council (FSOC) in 2010 to identify and monitor risks to the US financial system, promote market discipline, and respond to threats to the stability of the US financial system. FSOC also monitors climate-related financial risks to initiate the steps necessary to ensure the resilience of the financial system to such risks.

Given that the demand for information on climate-related risks has grown rapidly, FSOC is calling for high-quality disclosure of data by companies to improve information provision for investors and other market participants. The continued development and implementation of regulatory and financial frameworks will be critical to minimizing risks to financial systems. After all, they remain the necessary backbone of a broad-based initiative to address climate change. Extensive legislative action will be required to enable such disclosure by companies and other actors required by the FSOC.

Like all the other smaller banks, the lenders of last resort, the big banks, and the world banks see their primary responsibility as ensuring that their investments and assets perform positively. After all, we live in a capitalist system where profit comes first. Thus, so long as investments in ESG assets, companies, and businesses are profitable, most banks and funds, as well as other private investors, will jump on the ESG bandwagon and consequently finance the green revolution.

References

European Central Bank. https://www.ecb.europa.eu/home/html/index.en.html

International Monetary Fund. https://www.imf.org/en/Home

The World Bank. (n.d.). World Bank open data. https://data.worldbank.org/

US Department of the Treasury. (n.d.). Financial Stability Oversight Council. https://home.treasury.gov/policy-issues/financial-markets-financial-institutions-and-fiscal-service/fsoc

US Department of the Treasury. (2021). Remarks by Secretary Janet L. Yellen at the Open Session of the Meeting of the Financial Stability Oversight Council. October 21. https://home.treasury.gov/news/press-releases/jy0424

What is ESG?

E SG is the abbreviation for environmental, social, and governance. These criteria represent a set of still-to-be-defined standards for potential investors. Environmental means that companies must act in an environmentally friendly way. Social refers to how companies treat people (workers, employees, customers, suppliers, and the communities in which the company is located). Governance means how well a company's management, compensation, internal audits, controls, and relationships with shareholders are managed. ESG criteria are becoming increasingly important in evaluating companies in which investors put their money. In addition, a growing number of investors are interested in aligning their investments with their own climate goals, either because of their mission statement or a recognition that there can be no profit on a dead planet. Accordingly, providing information tools based on standardized, high-quality, and audited practices would help investors, stakeholders, employees, and the public alike achieve a faster transition to a sustainable world.

In addition, another fundamental problem exists. As one German banker noted, "The basic problem is that smaller companies often don't have the same resources to present their sustainability reporting in as much detail as large companies. But that doesn't mean that small caps are less sustainable."

In the future, it will be essential to determine a company's sustainability and provide accurate figures on its environmental impact, but what are the crucial sustainability issues? The big problem is that it remains unclear who will determine and verify the complete set of sustainability issues at the national and international levels. All of this is new territory for global markets, and yet ESG is only now maturing into an "effective and sustainable" investment strategy. In this context, a new concept has already been established, called "double materiality." Double materiality defines how information disclosed by companies both has an effect on the value of the company and the company's impact on the world or environment, as such. Material environmental impacts of a product or company can quickly turn into financial risks. It is critical to remember that most investors base their investment decisions on the principle of prudence, i.e., they do not want to incur more risk.

The ESG "score" of companies — what they are doing and what it means for the environment — is not easy to determine and even less easy to monitor. Because of growing market pressure, most companies are now touting their ESG strategies in consistently positive and flowery terms. Portfolios promise a "holistic view of sustainability" or big growth targets "based on a comprehensive review of our sustainability approach and the company's long-term goals and aspirations." Typically, many companies' glossy portfolios do not include trackable numbers but instead opt for detailed views and descriptions of the issue; they promise thorough reviews of overall strategy and unique positive changes to meet all ESG requirements. However, without truly meaningful data, investors are unable to make climate-related decisions. To address this data gap, financial regulators and supervisors such as the Task Force on Climate-related Financial Disclosures (TCFD) are in the process of defining precise ESG requirements. The TCFD was established by the Financial Stability Board (FSB) in 2015. It has 32 members from across the G20 group, an international forum of 19 countries and the European Union (EU).

However, there are other players working on ESG standards, such as the International Financial Reporting Standards (IFRS) Foundation. The Foundation introduced the International Sustainability Standards Board (ISSB) at COP26 in 2021. The aim of the ISSB — similar to the

TCFD — is to create a basis for corporate sustainability standards in order to provide the financial markets with all the ESG information they need. The ISSB is chaired by Emmanuel Faber, former CEO of Danone, a food company.

The Sustainability Accounting Standards Board (SASB) is another independent non-profit organization that has already established specific standards for companies to use when disclosing ESG information to investors. The standards provide detailed industry-specific topic sets to identify all issues relevant to financial performance in more than 70 industries. The Climate Disclosure Standards Board (CDSB) is an international consortium of companies and environmental non-governmental organizations; it provides a framework for companies to report their ESG information.

Given this landscape shift, almost all companies are in the process of developing ESG standards for the 21st century. Until these standards are universally defined, there is no reliable, comprehensive way to determine how "green" or "sustainable" a company really is. Sustainability reporting currently lacks a baseline of generally accepted standards that already exist in other areas of the financial world. Thus, it may take some time before there are international, uniform standards for recording sustainability criteria for companies.

Being the CEO of an ESG-certified company in the modern world is a very different ballgame from the usual process. You must address urgent global challenges while simultaneously rebuilding large parts of a new economy. As Marc Benioff, CEO of Salesforce, puts it, "Capitalism as we know it is dead, and that new kind of capitalism that's going to emerge is not the Milton Friedman capitalism that's just about making money. If your orientation is just about making money, I don't think you're going to hang out very long as a CEO or a founder of a company."

Of course, there are strong headwinds from the powerful fossil fuel industry. CEOs of fossil fuel companies see that their companies' assets could melt like ice in the sun. If their coal, oil, or gas stays in the ground, the companies will inevitably lose those assets. Phil King, a Texas state legislator, said, "Oil and gas is the lifeblood of the Texas economy." In France, the issue is a different panacea. In order to keep the French nuclear industry — 56 large nuclear reactors operate there — alive, the EU Commission has classified nuclear technology as "sustainable." In a market where the primary concern is keeping the economy afloat, compliance with sophisticated ESG standards may be secondary for companies. Besides, these global standards do not yet exist.

The Sustainable Finance Disclosure Regulation (SFDR) — a European regulation applicable since 2023 — states that investors must disclose how they incorporate sustainability risks into their decisions and how they report to final beneficiaries on their strategy, objectives, and impacts. We will have to wait and see what effect this policy will have in the foreseeable future.

References

Beerbaum, D.O. (2021). Green Quadriga? — EU — Taxonomy, TCFD, Non-Financial-Reporting Directive and EBA ESG Pillar III/ IFRS Foundation. Social Science Research Network, April 11. https://papers.ssrn.com/sol3/papers.cfm?abstract_id=3824397

Benioff, M. (2019). We need a new capitalism. *The New York Times*, October 14. https://www.nytimes.com/2019/10/14/opinion/benioff-salesforce-capitalism.html

BTG Pactual. (n.d.). ESG & impact investing. https://www.btgpactual.com/uk/esg-e-impact-investing

Climate Disclosure Standards Board. https://www.cdsb.net/

Douglas, E. (2021). Texas Legislature advances bills to shield oil and gas from climate initiatives. *Texas Tribune*, May 3. https://www.texastribune.org/2021/05/03/texas-house-fossil-fuel-oil-divest/

European Commission. (n.d.). Corporate sustainability reporting. https://finance.ec.europa.eu/capital-markets-union-and-financial-markets/company-reporting-and-auditing/company-reporting/corporate-sustainability-reporting_en

European Commission. (2021). Proposal for a directive of the European Parliament and of the Council amending Directive 2013/34/EU, Directive 2004/109/EC, Directive 2006/43/EC and Regulation (EU) No 537/2014, as regards corporate sustainability reporting. April 21. https://eur-lex.europa.eu/legal-content/EN/TXT/PDF/?uri=CELEX:52021PC0189&from=EN

Green Finance Institute. https://www.greenfinanceinstitute.com/

HM Treasury, Department for Work & Pensions, and Department for Business, Energy & Industrial Strategy. (2021). Greening finance: A roadmap to sustainable investing. October. https://assets.publishing.service.gov.uk/government/uploads/system/uploads/attachment_data/file/1031805/CCS0821102722-006_Green_Finance_Paper_2021_v6_Web_Accessible.pdf

IFRS Foundation. (2021). Constitution. https://www.ifrs.org/content/dam/ifrs/about-us/legal-and-governance/constitution-docs/ifrs-foundation-constitution-2021.pdf

Sarasin & Partners. https://sarasinandpartners.com/row/

Sustainability Accounting Standards Board. https://sasb.org/

Sustainability Accounting Standards Board. (n.d.). SASB standards. https://sasb.org/standards/download/

Task Force on Climate-related Financial Disclosures. https://www.fsb-tcfd.org/

Emissions Trading

In a 1960 essay, British economist Ronald Coase (1910–2013, Nobel Prize in Economics 1991) argued that parties in dispute over property rights — e.g., who owns CO2? — will find an optimal solution under the right conditions. In the years that followed, carbon emission models were created in several US cities. This allowed for a comparison of the costs and effectiveness of different control strategies for the rapidly increasing air pollution nationwide. In 1972, the US Environmental Protection Agency (EPA) introduced the concept of computer modeling with strategies to reduce costs (e.g., through emissions trading). This eventually led to the concept of emissions trading (cap and trade) as a means of achieving cost-effective solutions.

Cap and trade (CAT) regulations govern the amount of CO2 and other environmentally harmful gasses that a company is allowed to emit into the environment ("cap"). If the company emits more CO2, it must buy emission rights. On the other hand, a company can also sell unused emission rights ("trade"). This process creates an economic incentive to reduce GHG emissions. China, the world's largest emitter of GHGs, launched the first phase of its own national carbon market in 2017 with the support of the EDF (Environmental Defense Fund). Under the command-and-control (CAC) approach used by China, a central authority sets the levels of how much a manufacturing facility is allowed to emit.

An emissions trading system (ETS) was introduced in the EU in 2005. It was the first major GHG emissions trading program in the world. The ETS covers more than 11,000 power plants and factories, and the principle behind it is very simple. The goal is to reduce GHG emissions from factories and power plants by putting a price on those GHG emissions. Companies in the EU must buy and can trade European emission allowances (EUAs) for every ton of CO2 they emit within a calendar year. CO2 emissions trading in the EU reduced GHG emissions by about one gigaton between 2008 and 2016. The EU has set a new target for 2021. GHG emissions are to fall by at least 55% by 2030 compared to 1990 ("Fit for 55"). In 2021, Germany introduced an emissions trading system for almost all CO2 emissions resulting from the combustion of fossil fuels (Fuel Emissions Trading Act (BEHG)). This means that emissions prices will also be levied for transport and heating. The control obligations remain with the companies that provide fossil fuels rather than with individual motorists or homeowners.

All existing trading models are criticized by environmental initiatives, some of them very strongly. They believe that the increasingly intense effects of climate change that can be observed require far more radical action than the current emissions trading systems. Long-term structural change will not be possible if the existing structures of emissions trading and the relatively favorable CO2 prices that have prevailed to date continue. Many of the emissions premiums offered on the global market come from less developed countries; their purchase by large CO2 emitters means that potential GHG reductions can be sidestepped. In addition, there are too many different standards and systems (similar to ESG "standards"). What we need is a simple, understandable CO2 pricing system on a worldwide scale.

References

Environmental Defense Fund. (n.d.). How cap and trade works. https://www.edf.org/climate/how-cap-and-trade-works

European Commission. (n.d.). Emissions cap and allowances. https://climate.ec.europa.eu/eu-action/eu-emissions-trading-system-eu-ets/emissions-cap-and-allowances_en

European Commission. (n.d.). 'Fit for 55': Delivering the EU's 2030 Climate Target on the way to climate neutrality. https://eur-lex.europa.eu/legal-content/EN/TXT/HTML/?uri=CELEX%3A52021DC0550

Federal Ministry for the Environment, Nature Conservation, Nuclear Safety and Consumer Protection. (n.d.). Act on National Allowance Trading for Fuel Emissions Fuel Emission Allowance Trading Act. https://www.bmuv.de/fileadmin/Daten_BMU/Download_PDF/Gesetze/behg_en_bf.pdf

Wall Street Mojo. (n.d.). The Coase theorem. http://coase-theorem/

Other Trading Models

C hatham House is a private think tank focused on international affairs. The main goal of this internationally renowned institution is "to help governments and societies build a sustainably secure, prosperous, and just world." The goal of the so-called "Chatham House Sustainability Accelerator" is to channel more capital into nature-based solutions that benefit both the environment and society. We absolutely need more such ideas today and in the future — ideas that can then be quickly implemented by market players. Only with such instruments can important technologies for combating climate change be implemented quickly and in a targeted manner.

At a lower level, an initiative group calling itself "Tomorrow's Air" and advertising itself as a "community of courageous travelers" has recently teamed up with Climeworks, a Swiss "direct air capture" (DAC) company. The goal is to help "passionate travelers" reduce their environmental impact and thus "continue to make the joys and benefits of (air) travel possible." Climeworks currently sells carbon offsets for 980 euros per ton of CO_2 — about as much as is emitted per seat on a round-trip flight from London to New York. The message of this initiative for air travelers is that a sensible climate policy can be based on offsetting GHGs.

In contrast, "Don't fly" is part of a strategy aimed at limiting, if not eliminating, air travel. Another step in this direction is reducing red meat consumption and meat consumption in general. For some people, of course, cheap airplanes and beef steaks are an essential part of their lifestyle. Many of us, especially in the developed world, have become accustomed to these things. For a different, greener, and more resilient world, all of this must be abandoned or at least reduced to healthier levels.

In Finland, an organization called Puro has launched the world's first business-to-business (B2B) marketplace focused exclusively on carbon removal. Their business model is to offer CO_2 removal as a service. Puro makes it possible to neutralize emissions by investing in "negative concrete," for example. CarbiCrete is a company that already offers such "negative concrete." In the production process, cement is replaced by steel slag, thus avoiding CO_2 emissions.

In the US, the start-up company Nori has attracted a lot of attention because it sells carbon credits from soil sequestration under the slogan "Remove Carbon to Help Reverse Climate Change." The company promises to "remove your carbon footprint for a period of time you choose." Those with a particularly guilty conscience can pay off their carbon "debt" retroactively for the past five years. Nori's business model may just be another way for the rich to offset their emissions, the latest in a long line of increasingly creative ways to avoid the aforementioned and dreaded "abstinence" from CO_2-intensive activities and pleasures — in other words, greenwashing. Instead, avoid flying, and please sell your gas-guzzling SUV. Change your life!

References

CarbiCrete. https://carbicrete.com/

CBS International Business School. (2021). Sustainable living: 58 tips for a more sustainable lifestyle. May 26. https://www.cbs.de/en/blog/sustainable-living-tips-for-a-more-sustainable-lifestyle/

Chatham House. https://www.chathamhouse.org/

Nori. https://nori.com/

Puro.earth. https://puro.earth/

The World Bank. (n.d.). Carbon pricing dashboard. https://carbonpricingdashboard.worldbank.org/what-carbon-pricing

Tomorrow's Air. https://www.tomorrowsair.com/

What is Capitalism?

The essential characteristic of capitalism is its profit orientation. Profit plays a decisive role for people who invest. Investors want their money to work for them; they expect a financial reward for the risk of their investments. Capitalists expect profits.

Capitalism is based on several pillars:

- Private property — People are allowed to own (a) tangible assets such as land and houses, and (b) tangible assets such as bonds or stocks, etc.
- Self-interest — People act in their own interest without political or social pressure. People's actions ultimately benefit society as a whole, as if guided by an invisible hand (Adam Smith).
- Competition — Companies have the freedom to both enter and exit markets in order to maximize societal welfare.
- Fixed functioning market mechanism — Prices are determined only by interactions between sellers and buyers, who naturally want to obtain either the highest or lowest price for goods, services, and wages.
- Freedom — People have the freedom to choose how they produce, consume, and invest. Dissatisfied customers can buy other products, investors can look for more lucrative ventures, and workers can look for other, better-paying jobs.
- Limited role of the government — The state is constrained in protecting citizens' rights and maintaining an adequate environment for markets to function smoothly.

This is a theory of ideal capitalism. In real-world capitalism, as well as real-world socialism, where the state owns all means of production, there is one problem — the people.

Figure 17.   Adam Smith (1723–1790). (Public Domain)

Adam Smith, the founding father of classical economics, characterized the capitalist man according to what he saw in his time and what remains true today: They are men who generally have an interest in deceiving and even oppressing the general public and who have, therefore, both deceived and oppressed the public on many occasions. This rather gloomy view of capitalists has lost none of its veracity, especially before (and after) the global financial crisis of 2007/2008, when excessive risk-taking by global financial institutions, insane lending practices to low-income homebuyers, the tying of real estate to mortgage-based securities, and a vast network of financial derivatives built by clever traders led to the near-collapse of the global financial system.

The ideas of economists and political philosophers, both when they are right and when they are wrong, are more powerful than is commonly understood. Indeed, the world is ruled by little else.

Milton Friedman (1912–2006) was an American economist who received the Nobel Prize in Economics in 1976. With his market interpretations, he became the main opponent of Keynesian theory. The Briton John Maynard Keynes (1883–1946) developed his economic theory in order to avoid major crises, such as the Great Depression of the 1930s, in the future. In *The General Theory of Employment, Interest and Money* (1936), Keynes advocated higher government spending and lower taxes, i.e., active stabilization methods and economic policy intervention by governments.

In contrast, Friedman argued that the opposite was true. Friedman denounced any "social responsibility of business in a free market system." Corporate leaders who speak and act with a "social conscience" are, in his eyes, preaching pure socialism and thus undermining the foundations of a free society. According to Friedman, business leaders are merely puppets of business owners who have only one goal — profit. Friedman, in an essay for *The New York Times* on September 13, 1970, explains, "There is one and only one social responsibility of business — to use its resources and engage in activities designed to increase its profits so long as it stays within the rules of the game, which is to say, engages in open and free competition without deception or fraud."

Economic growth in the Global North is largely dependent on low incomes and cheap resources in the Global South. As the countries of the Global South became more independent after World War II, they began to raise wages while also trying to regain control over their resources. In the 1970s, for example, this led to the so-called "oil price shock." In order to maintain capital accumulation, the rich countries of the North shifted to what became known as "Neoliberalism" — the dominant political and economic philosophy of recent decades.

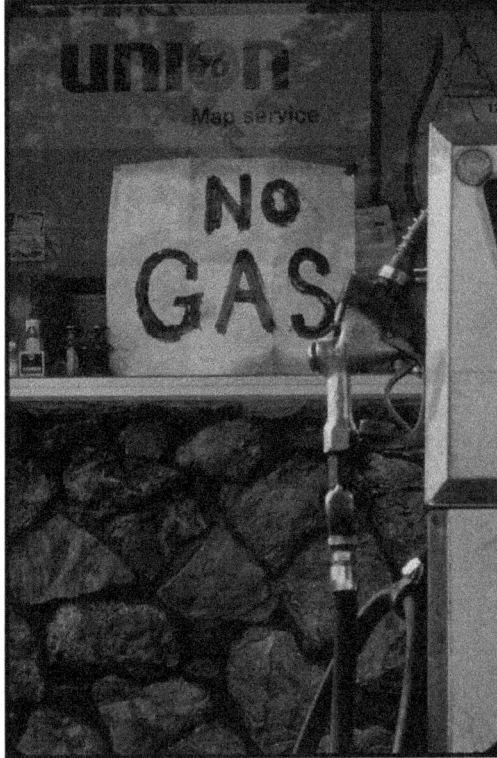

Figure 18. Oil price shock, US, 1973. (Public Domain)

Modern capitalism, mixed economy, free play of market forces, the theory of neoliberalism, and market-compliant policies (as characterized by German ex-Chancellor Angela Merkel) are just a few catchy buzzwords about capitalism that have arisen since the unsettling shock of 2007/2008. How will capitalism evolve in the coming times of radical global challenges when a truly solid financial backbone will be essential for building safe, globally functioning alternatives to the fossil fuel-based economy? How will the so-called "Piketty effect" — the fact that wealth is growing faster than income — play out in the future, given the record stock market values and escalating social problems?

Wolfgang Streeck (b. 1946), a German social scientist, is director emeritus at the Max Planck Institute for the Study of Societies in Cologne. He has a rather startling answer to the question of how capitalism will develop in the coming period: "In my opinion, it is high time ... to rethink capitalism as a historical phenomenon that not only has a beginning, but also an end."

Streeck characterizes the era of capitalism after its "marriage" to democracy as the story of a long and painful period of cumulative decline. Growth gives way to secular stagnation, inequality leads to instability, and confidence in the capitalist monetary economy has all but disappeared. There are at least five worsening disorders of postwar capitalism for which there is no cure: declining growth, oligarchy, emaciation of the public sphere, corruption, and international anarchy. According to Streeck, no one believes in a moral revival of capitalism anymore.

Streeck argues that we are not only witnessing the decline of neoliberalism but also the final phase of global capitalism. However, he has no answer to the question of what comes next. At the same time, he is aware that the end of capitalism is a byproduct of the doctrine itself: "In fact, since the term became common in the mid-18th century, all the major theorists of capitalism have predicted its imminent demise. This is true not only of radical critics like Marx or Polanyi but also of bourgeois theorists like Ricardo, Weber, Schumpeter, Sombart, and Keynes."

After the neoliberal revolution, capitalism is facing its end without a clear alternative or even a prospect of what will come after. Like climate change, capitalism is evolving into something very different from what it was before. Quite simply, humanity has been too successful; we have eaten the earth.

Capitalism has also been too successful, undermining itself with no successor or even a viable alternative in sight. Economic growth, social balance, and financial stability are no longer guaranteed. The whole system seems to be falling further and further apart with no helping hand in sight. This will not mean the end of the world but merely the disappearance of one feature — namely, capitalism as a predictable, just, and viable social order.

References

Baller, S. (n.d.). How the economic machine works a template for understanding what is happening now Ray Dalio Bridgewater. Academia.

Cohen-Setton, J. (2015). The Piketty theory controversy. Bruegel, April 14. https://www.bruegel.org/blog-post/piketty-theory-controversy

Dalio, R. (n.d.). Princples. http://xqdoc.imedao.com/153ac69641721b3faf7ab545.pdf. https://www.britannica.com/money/topic/financial-crisis-of-2007-2008

Encyclopaedia Britannica. (n.d.). Financial crisis of 2007–08. https://www.britannica.com/money/topic/financial-crisis-of-2007-2008

Encyclopaedia Britannica. (n.d.). Neoliberalism. https://www.britannica.com/money/topic/neoliberalism

Friedman, M. (1970). The social responsibility of business is to increase its profits. *The New York Times*, September 13. https://www.nytimes.com/1970/09/13/archives/a-friedman-doctrine-the-social-responsibility-of-business-is-to.html

Keynes, J.M. (1936). Concluding notes on the social philosophy towards which the General Theory might lead, in *The General Theory of Employment, Interest and Money*. https://www.marxists.org/reference/subject/economics/keynes/general-theory/ch24.htm

Max Planck Institute for the Study of Societies. (n.d.). Prof. em. Dr. Dr. h.c. Wolfgang Streeck. https://www.mpifg.de/457994/Streeck

Merle, R. (2018). A guide to the financial crisis — 10 years later. *Washington Post*, September 10. https://www.washingtonpost.com/business/economy/a-guide-to-the-financial-crisis--10-years-later/2018/09/10/114b76ba-af10-11e8-a20b-5f4f84429666_story.html

Stanford Encyclopedia of Philosophy. (2021). Neoliberalism. June 9. https://plato.stanford.edu/entries/neoliberalism/

Streek, W. (n.d.). How will capitalism end? *New Left Review*. https://newleftreview.org/issues/ii87/articles/wolfgang-streeck-how-will-capitalism-end

Philanthropy

According to the Cambridge Dictionary, philanthropy means:

- helping the poor, especially by giving them money (in the UK).

- the giving of money, especially in large amounts, to organizations that help people (in the US).

- derived from the Greek words "philos" ("to love") and "anthropos" ("human being").

Bill Gates, Warren Buffett, MacKenzie Scott (ex-wife of Jeff Bezos), and several other prominent individuals set a new standard for "public accountability for charity" when they pledged in August 2010 to donate the majority of their wealth to charity. In Bill Gates' words, "This is about building on a wonderful tradition of philanthropy that will ultimately help the world become a much better place." By 2022, there were a total of 231 "Pledgers," including Michael Bloomberg, Elon Musk, Mark Zuckerberg, Richard Branson, and Hasso Plattner (SAP), rounding out some of the more well-known figures.

The Giving Pledge is based on a simple formula: The charitable campaign is open to billionaires who are willing to give most of their wealth to philanthropy either during their lifetime or in their wills. In his letter establishing the Pledge, Warren Buffett, considered the most successful investor of the 20th century, wrote: "99% of my wealth will go to philanthropy during my lifetime or at death. 99% can have a huge effect on the health and welfare of others." Buffett further offered insight into the system that brought him enormous wealth in the first place, which reflects his strong insight into the core of his success: capitalism — "My

Figure 19.   Elon Musk, "Pledger." (Source: Debbie Rowe, CC BY-SA 4.0, https://commons. wikimedia.org/w/index.php?curid=125967789)

luck was accentuated by my living in a market system that sometimes produces distorted results, though overall it serves our country well. I've worked in an economy that rewards someone who saves the lives of others on a battlefield with a medal, rewards a great teacher with thank-you notes from parents, but rewards those who can detect the mispricing of securities with sums reaching into the billions. In short, fate's distribution of long straws is wildly capricious."

There is a curious problem with the Pledger's fortune: the pace of their growth. Pledgers' wealth has grown by almost 100% in the last 10 years, and if they are to keep their promise to give away most of their money, they must spend their money faster. This development dramatically illustrates the imbalance of wealth growth in the modern neoliberal financial system and also how difficult it is to spend money intelligently.

Ray Dalio, a hedge fund manager, joined The Giving Pledge in 2011. Dalio, sometimes referred to as the "Steve Jobs of investing," is one of the world's most successful investors, with a net worth of $18 billion. His company, Bridgewater, manages $160 billion. Dalio sees a crisis coming in society because the economic gap in society is widening so rapidly. He describes the situation as unjust, unproductive, and threatening; the concentration of wealth in fewer and fewer hands is a danger to democracy. For this reason, he established a foundation, "The Dalio Foundation," as his personal funding vehicle. It supported the TED project "Audacious," which funds socially minded entrepreneurs working to solve global problems, and it has just pledged a record $100 million to create learning environments that will engage students and transform the educational experience and outcomes in Connecticut schools. Dalio says, "Our goal is to contribute to equal healthcare and equal education because we believe that these are the most fundamental building blocks of equal opportunity and a just society." Dalio's main focus as a philanthropist is on the world's oceans because the oceans are the world's largest and most important natural resource. The ocean is needlessly neglected; less than 5% of it has been explored.

Chuck Feeney has chosen a different, more active path. He co-founded the retail giant Duty Free Shoppers and has consistently implemented his idea of "Giving While Living" by donating more than $8 billion to charities, universities, and foundations around the world through his foundation, Atlantic Philanthropies. He kept about $2 million for himself and his wife while distributing his entire $8 billion anonymously, which is why Forbes magazine called him the "James Bond of Philanthropy." Feeney spent most of the money, $3.7 billion, on investments in the school system. One billion of that went to his former university, Cornell. He donated more than $870 million to organizations working for human rights and social improvement. He spent $62 million on an effort — thus far unsuccessful — to abolish the death penalty in the US. Feeney described his motivations plainly: "I see little reason to delay giving when so much good can be achieved through supporting worthwhile causes. Besides, it's a lot more fun to give while you live than give while you're dead." Warren Buffett commented, "Chuck is a role model for all of us."

Jeff Bezos, currently the second-richest person in the world, has earmarked $10 billion for his "Earth Fund" to combat climate change. The timetable is tight: the entire $10 billion must be spent within a decade. The fund is designed to demonstrate the need for data- and science-based change. It focuses on the dynamic, positive impacts of ambitious climate action triggered by new technologies, increased economic efficiency, changing expectations for the future, benefits for first movers, and entrepreneurial thinking.

Since its launch in 2020, Bezos' Earth Fund has poured serious money into large, well-known environmental groups, including the World Resources Institute, the Environmental Defense Fund, the Rocky Mountain Institute, the ClimateWorks Foundation, and the Natural Resources Defense Council. This first round of Earth Fund awards was met with criticism because some of the grantees — the Nature Conservancy, for example — already have large market capital, whereas smaller institutions were simply "overlooked." One of the smaller grant recipients is The Solutions Project, which received $43 million. The project invests specifically in the global movement for climate justice, centered on women and organizations led by indigenous people, immigrants, and "people of color."

In December 2021, Bezos announced the next round of grants totaling $443 million from the Earth Fund. Recipients will include programs focused on climate justice, conservation, and restoration, with a focus on Africa and the South American continent. This round of funding is intended to create change in these areas that can serve as a model for others to see what needs to be done in the future. Earth Fund CEO Andrew Steer, a longtime World Bank employee, announced that the Bezos Earth Fund will also invest in the private sector. This announcement suggests that some of the Earth Fund's spending could be used as an investment engine.

Bezos' megacorporation Amazon is taking other paths toward improvement: the company's CO_2 emissions continue to rise, even though Bezos has promised that the company will achieve net-zero emissions by 2040. Nevertheless, Amazon is pushing ahead with change: the company has already ordered 100,000 electric delivery trucks to be manufactured by Rivian, a US start-up. Another Bezos venture, Blue Origin, is an aerospace company that has already announced plans to build its own space station. The project, called Orbital Reef, is scheduled to launch in 2025–2030 and will be a kind of commercial space area. On the one hand, Bezos is donating a lot of his own money to the Earth Fund, but on the other hand, his space program, with its many rocket launches, will be a significant polluter.

References

Bertoni, S. (2020). The billionaire who wanted to die broke . . . is now officially broke. Forbes, September 15. https://www.forbes.com/sites/stevenbertoni/2020/09/15/exclusive-the-billionaire-who-wanted-to-die-brokeis-now-officially-broke/?sh=1fc086f13a2a

Bezos Earth Fund. https://www.bezosearthfund.org/

Blue Origin. https://www.blueorigin.com/

Bridgewater. https://www.bridgewater.com/

Dalio, R. (2011). Principles. http://xqdoc.imedao.com/153ac69641721b3faf7ab545.pdf

Rivian. https://rivian.com/

The Atlantic Philanthropies. https://www.atlanticphilanthropies.org/

The Giving Pledge. https://givingpledge.org/

The Solutions Project. https://thesolutionsproject.org/

Taxes

W hy not just tax the rich, as in ancient Greek democracy? Changes to the tax systems of the Western world, particularly in the US, have led to a massive drop in taxes for the rich in recent years. Elon Musk, the richest person in the world, with an estimated fortune of $270 billion, reportedly did not pay a dime in income tax in his home country, the US, in 2020. Focused on this phenomenon, Emmanuel Saez, a French economist, and Gabriel Zucman, a professor at UC Berkeley, have published the book *The Triumph of Injustice*. It is a careful study of tax trends for different social groups from the early 20th century to the present. According to this book, "the rich" — people with annual incomes of more than $1.5 million — now pay less tax in the US than other social groups (23%) compared to 25–30% tax burdens among the working and middle classes. The authors call this regressive taxation "a new engine of inequality."

The US has very regressive income taxes, where the tax burden declines with income. In contrast, payroll taxes are rising and minimum wages are low, but income taxes on capital and capital gains are also low. Corporate taxation almost halved in 2018 as compared to 2017. Another crucial trend is reflected in the fact that real income for the bottom 50% of people in the US has stagnated since 1980 while the rich are getting richer. The authors refer to this trend at the top end of the scale as "snowballing wealth accumulation." Among the top 1% of the rich, annual income has risen significantly, from $5 trillion in 1982 to $33 trillion in 2018.

In order to create a new and fairer tax system that counteracts the current direction, the book's authors propose to end the rampant corporate tax evasion that was made public by the "Panama Papers." They recommend:

- A 25% tax on corporations on a country-by-country basis and a wealth tax on the truly rich with more than $1 billion in assets
- Progressive taxation of higher incomes, as was the practice until 1980
- A new system of income taxation, which they call "reinventing the income tax," that takes into account the need to fund health care and better education.

References

Kagan, J. (2023). Regressive tax: Definition and types of taxes that are regressive. Investopedia, March 18. https://www.investopedia.com/terms/r/regressivetax.asp

Saez, E. and Zucman, G. (n.d.). The triumph of injustice (W.W. Norton). https://www.oecd.org/naec/events/multidimensional-well-being/G_Zucman.pdf

Tax Justice Now. https://taxjusticenow.org/

Think Tanks and Endowments

C limateWorks is a US-based non-profit foundation established in 2008. Its mission is to limit global warming by funding other organizations internationally. ClimateWorks focuses on energy, industry, buildings, transportation, and forestry. The foundation is funded by major philanthropic players such as Oak, Packard, and MacArthur and is one of the 100 largest charities in the US. ClimateWorks is a major player in the field, with an annual budget of about $200 million. The founder and first CEO of the ClimateWorks Foundation was Hal Harvey, now CEO of Climate One, a platform that addresses climate change by bringing together private, social, and governance decision-makers.

ClimateWorks cooperates with a large network of universities, institutions and departments. In their policy paper, "Climate Risks," they look more closely at near-term (through 2030) transition risks and longer-term (through 2050) physical climate risks, i.e., the problem of "too much and too little." Transition risks imply that certain parts of the economy could face major transitions or higher costs — such as higher carbon prices — as a result of impending global change. Physical risks refer to severe weather events, such as devastating floods, extreme storms, wildfires, or droughts — in the words of David Pogue, author and CBS host: the threat of climate chaos or even climate collapse.

Transition risks are, for example, lost assets such as decommissioned coal- or gas-fired power plants or decommissioned nuclear power plants. These actions mean the loss of money invested in these assets and lost revenue when plants that are still in operation are shut down. Physical risks like droughts, heat waves, and floods will increase in the more distant future (2050) if the necessary countermeasures are taken too late. The world is already on this path. It is estimated that the number of heat waves will increase fivefold by 2050 under a 1.5°C scenario and tenfold under a no-countermeasures scenario. Agricultural droughts will follow a similar trajectory.

Agora Energiewende, a German think tank and part of the non-profit Smart Energy for Europe Platform (SEFEP), develops viable strategies to promote climate neutrality (Net Zero). Agora was founded in 2012 and has become one of Europe's leading think tanks on energy and climate policy. The organization makes informed proposals to policymakers and the public and is building an international network of similar partner organizations. Agora promotes continuous dialogue and intensive discussions between stakeholders in politics, business, and civil society.

The key question behind Agora's activities is how to achieve a rapid transition of the entire economy toward Net Zero. With a budget of under eight million euros, Agora is rather small when compared to large organizations like ClimateWorks. However, Germany and Europe as a whole have a very different organizational network, and German governments, especially with the Green Party as part of the current government, are putting the path to a Net Zero economy and the country as a whole at the center of their political agenda.

Chatham House (see p. 95) is a think tank founded in 1920. The organization's goal is to help governments and societies build a sustainable future across a wide range of issues, including politics, economics, defense, health, and technology, among others. The Chatham House Rule is as follows: When a meeting or portion thereof is held under the Chatham House Rule, participants are free to use the information they receive but neither the identity nor affiliation of the speaker(s) nor that of any other participant may be disclosed. Online events they hold are open to the public.

London-based Chatham House offers some 300 events each year — conferences, workshops, and roundtable discussions — all of which serve the House's mission to be "a world-leading source of independent analysis, informed debate, and influential ideas on how to build a prosperous and secure world for all." Chatham provides in-depth analysis, hosts high-profile speakers from around the world, and organizes conferences and events both in the UK and overseas.

References

Agora Energiewende. https://www.agora-energiewende.org/

Chatham House. https://www.chathamhouse.org/

Climateworks Foundation. (n.d.). Climate risks. https://www.climateworks.org/wp-content/uploads/2021/12/Climate-Risk-Companion-Deck_ClimateWorks-12.14.21.pdf

YouTube. (2020). How to decarbonize the grid and electrify everything | John Doerr and Hal Harvey. November 20. https://www.youtube.com/watch?v=d4Cy16uOdLM

"Green" Banking

The German bank GLS was founded in 1974. By its own account, it is the world's first eco-bank. Its business model is based on social and ecological criteria that shape its entire investment and financing business. Both exclusion and positive criteria apply to all loans and to the securities and investment business. For the bank, positive criteria include those projects and companies that pursue sustainable and future-oriented goals. The bank guarantees it will not invest in any of the following business areas, which are exclusion criteria: nuclear energy, carbon energy, armaments and weapons, biocides and pesticides, genetic engineering in agriculture, chlororganic mass products, factory farming, embryo research, or addictive substances.

Companies, products, and services in which investments are made should operate exclusively in future-oriented, social-ecological business sectors. These include pioneering companies for sustainable business that develop sensible and sustainable solutions, renewable energies, and production, and processing and trade of and with agricultural products and healthy foodstuffs. In construction financing, the bank's focus is on energy-efficient buildings with a positive energy balance or the lowest possible primary energy consumption with reasonable life cycle costs. In the areas of education and culture, facilities are financed that promote the personal development of children and young people on the basis of a basic liberal approach. In the social and health sector, the bank focuses on non-profit facilities and private sponsors, e.g., holistic care facilities such as nursing homes and assisted living facilities for people with disabilities, child and youth welfare, or people with mental illness, as well as hospitals, doctors' practices, and health-promoting facilities.

Other examples of projects rated positively by the bank include environmentally and resource-friendly mobility systems, such as local public transport, the shift of freight traffic from airplanes or trucks to rail, and all aspects of cycling as a transport alternative. Within the sustainable economy, they fund forward-looking, social-ecological business areas, as well as sustainable commercial real estate, sustainable construction, recycling, eco-textiles, and natural cosmetics. Other criteria include good corporate governance, consistent alignment of business activities to the needs of employees, the community and society, and resource-conserving corporate management. Products and services that contribute to solving current social and ecological challenges are also rated positively. GLS Bank thus tries to apply ESG criteria in its portfolio and thereby ensure financing in the private to medium-sized sector that may not (yet) be guaranteed by major banks. This will change in the near future, as the major banks are in the process of adopting ESG criteria in a comprehensive manner.

Reference

GLS Bank. https://www.gls.de/privatkunden/english-portrait/

Well-Being

Tim Jackson (b. 1957) is professor of sustainable development at
the University of Surrey and Director of the Centre for the
Understanding of Sustainable Prosperity (CUSP). Professor
Jackson challenges what a business should really be doing. Is business about
producing as much as possible and throwing it away, thereby maximizing
shareholder profits? Or is the economy about creating wealth and well-being
for the citizens of a country? Shifting the focus from production to well-
being directs economic activity to areas such as care, health, social care,
education, renovation, and redevelopment, i.e., the areas that improve our
quality of life, employ an increasing number of people, and are proven
low-carbon sectors of the economy. The reason this has yet to happen is
that economists are committed to the idea of growth in productivity.
Current growth is mainly in the material-based sectors, where you can
replace people with machines and thus reduce the amount of work. That
is the wrong direction. This path leads to more unemployment and a more
material- and energy-intensive economy, and it does not address the gaps
in care and social welfare, nor those in our health care system and in the
construction of hospitals — in other words, all the things that could create
a better quality of life for all citizens. The transition to Net Zero should
not just be about capital markets putting a lot of money into new
technologies and machines but about transitioning to a decent economy
that has the well-being of all people at its core and supports the sectors that
support that well-being.

A recent study funded by the European Research Council (ERC) has shown that societies with low monetization achieve high levels of subjective well-being ("happy without money"). In such societies, people produce enough to satisfy their own needs and exchange only small amounts of non-essential goods and services.

The study was conducted in coastal communities in the Solomon Islands and in Bangladesh. The Solomon Islands is one of the largest island nations in the Pacific Ocean, and Bangladesh is one of the most densely populated countries in the world. Participants were asked to rate their happiness with life (which was asked as the first question of the survey to avoid participant bias) and to answer affective questions about the past day, i.e., how they felt about a selection of positive and negative events.

According to the study, high levels of satisfaction can be achieved even with very low levels of money in circulation. Overall, the study's findings challenge the prevailing view that economic growth is a reliable way to increase subjective well-being. Rather, it is culture that plays an important role in how happiness is perceived and conceptualized. The findings provide an unusually clear underpinning to the oft-discussed importance of non-material and non-market determinants of subjective well-being, and they raise reasonable doubts about the income-happiness debate as a whole.

In what are known as subsistence sites, people live their daily lives in close contact with nature, often embedded in a biologically diverse ecosystem. Recent work suggests that people generally tend to spend time in nature to enhance their well-being, with even stronger effects in pristine and biodiverse environments and when physically active in nature. A supportive and cohesive social network, or high social capital, is also widely recognized as a universal factor in subjective well-being. Traditional subsistence practices have been shown to strengthen solidarity with one another, as the entire community contributes to these activities, keeping the community connected to cultural traditions and its elders.

Economic growth is often seen as an essential step toward improving human well-being in developing countries. However, the findings of the ERC study mentioned above suggest that high levels of subjective well-being can also be achieved through the satisfaction of basic needs, access to a healthy natural environment, and social cohesion. The results also refute the view, based on the relationship between income and happiness, that sustainability is incompatible with high levels of happiness. A high level of subjective well-being in a society can, in fact, be achieved even with modest material spending.

Our economy is based on capitalism, which is fundamentally defined by its need to perpetually grow. However, we all know that the earth is not infinite; resources are getting scarcer, and the current system is ruining our lives and the planet itself. Thus, climate change can be seen as an expression of a failed concept — capitalism. Putting a lot of money into renewable energy and sustainable business is a good thing, but the main problem remains — growth. Remember the 1972 report about exponential growth in a finite world? It is clear that we have reached a fundamental turning point in our economic history. There are natural limits to our civilization. Well-being, a green economy, and less growth are the next important steps on our way to a functioning net-zero society.

In an interview, Professor Jackson was asked how we can maintain hope in the face of climate change. After a pause, Jackson recited a poem by Emily Dickinson and then said that something in the human soul ensures that hope will not fail us. The antidote to despair is not hope but action.

"Hope" is the thing with feathers —
That perches in the soul —
And sings the tune without the words —
And never stops — at all —

— "Hope", in: *The Poems of Emily Dickinson,*
edited by R. W. Franklin (Dickinson, 1999)

WEAll (Wellbeing Economy Alliance) is a collaboration of organizations, alliances, movements, and individuals working to transform the economic system. In an economy of well-being (the "new way"), people would be at the center of responses to climate change that focus on environmental protection and long-term sustainable regeneration. The exciting thing is that the new way is already emerging, with inspiring examples around the world showing us the way. As always, change begins with a desire, and only then will awareness emerge.

"We the people have the ability to solve all these problems facing us, if we just get about it."

— Hunter Lovins (b. 1950), American environmentalist,
co-author of *Natural Capitalism*

References

Dickinson, E. (1999). Hope. In *The Poems of Emily Dickinson*, Franklin, R.W. (ed.). Harvard University Press.

Miñarro, S., Reyes-García, V., Aswani, S, *et al.* (2021). Happy without money: Minimally monetized societies can exhibit high subjective well-being. *PLoS One*, **16**(1): e0244569.

Wellbeing Economy Alliance. https://weall.org/

YouTube. (2018). FUTUREMAKERS: Hunter Lovins. October 25. https://www.youtube.com/watch?v=81SEhfuFiV4

Post-Growth (Degrowth)

"Degrowth is about demolishing the imperial arrangement."

— Jason Hickel, Economic Anthropologist, 2022

Degrowth (or post-growth) is the key concept challenging the capitalist credo of steady and endless growth. Instead of increasing corporate profits, degrowth focuses on social and environmental well-being, on an end to overproduction and overconsumption. To achieve these goals, the entire economy must be transformed radically; shared values must shift from "bigger is better" to social care and solidarity. Degrowth respects planetary boundaries, ensures environmental justice, and aims for a complete and radical transformation of our society.

We need to leave our Growth Bubble, with its temptations and fleeting luxuries, to realize that growth cannot be the indispensable normality of our lives. The idea of a post-growth economy is not new. John Stuart Mill (1806–1873), one of the founding fathers of economics, already had ideas for a society after the one he lived in — the age of the Industrial Revolution. He particularly disliked "that the normal state of human beings is that of struggling to get on; that the trampling, crushing, elbowing, and treading on each other's heels, which form the existing type of social life, are the most desirable lot of human beings."

About 150 years after Mill, a young American politician decided to run for the Presidency of the United States, even though his older brother John had been assassinated in office: Robert Kennedy. On March 18, 1968,

Robert Kennedy gave a speech at the University of Kansas in which he addressed the war in Vietnam, poverty in the US, and his thoughts on why the gross national product (GNP) was an inadequate measure of success: "Too much and for too long, we seemed to have surrendered personal excellence and community values in the mere accumulation of material things. Our Gross National Product ... counts air pollution and cigarette advertising, and ambulances to clear our highways of carnage ... it counts the destruction of the redwood and the loss of our natural wonder in chaotic sprawl. It counts napalm and counts nuclear warheads and armored cars for the police to fight the riots in our cities ... it measures everything, in short, except that which makes life worthwhile. And it can tell us everything about America except why we are proud that we are Americans. If this is true here at home, so it is true elsewhere in the world." (YouTube, 2021).

These are grand statements from an election campaign that ended with the assassination of the candidate on June 4, 1968. But they demonstrate that today, we should think anew about what is really important in our lives, what really counts, and what stands above money and all the other things that can be counted. It is human dignity and what makes life worth living: happiness, togetherness, freedom, and love. Robert Kennedy's speech reflects the human desire for social justice in a cold and dangerous modern world, which is why it remains relevant today.

In the current economic system, economic growth increasingly favors only a few super-rich people. At the same time, the quality of life of most people is stagnating or even declining. Getting off the deadly treadmill would mean slowing down the whole hectic system, producing and consuming less as well as working less, to become a happier person in a healthier environment. We have a chance today to make our economy work for people and not for profit, so that more of us can live well.

The constant growth of our economies has already done so much damage that this process will eventually be irreversible. Once a species is extinct, it is gone forever. Once a resource is fully exploited, the supply will permanently disappear. We cannot get richer and richer while running away from the coming disasters because the disasters are already here. Compared to the existing system of technology-driven growth, degrowth scenarios minimize many key risks. In this context, the political and economic feasibility of change certainly depends on the will and skills of those in charge of business and politics. But in democracies, these individuals are legitimized by the people through elections. This means that we can help steer the change to a different orientation of the economic system through elections.

Economic anthropologist Jason Hickel writes about degrowth as part of a future structure of global justice. The main features of Hickel's "degrowth" scenario are:

- reduction of production and consumption
- reductions in working hours
- a fair distribution of income
- climate jobs guarantee
- a basic income
- universal public services

References

Brilhante, A.A. (2007). The centrality of accountability in John Stuart Mill's liberal-utilitarian conception of democracy. Ph.D. thesis, School of Public Policy, University College London, May. https://discovery.ucl.ac.uk/id/eprint/1444009/1/U591281.pdf

Degrowth. (n.d.). What is degrowth? https://degrowth.info/degrowth

Hickel, J. (2019). Degrowth: a theory of radical abundance. *Real-world Economics Review*, **87**. https://static1.squarespace.com/static/59bc0e610abd04bd1e067ccc/t/5cb6db356e9a7f14e5322a62/1555487546989/Hickel+-+Degrowth%2C+A+Theory+of+Radical+Abundance.pdf

Hickel, J. (2020). Degrowth: A response to Branko Milanovic. October 27. https://www.jasonhickel.org/blog/tag/degrowth

Hickel, J. (2022). Degrowth is about global justice: An interview. The Jus Semper Global Alliance. https://jussemper.org/Resources/Economic%20Data/Resources/JHickel-DegrowthGlobalJustice.pdf

Imboden, C. (2019). Write sign up sign in the revolution of hope: The practical philosophy of Erich Fromm. *Medium*, May 23. https://medium.com/swlh/the-revolution-of-hope-the-practical-philosophy-of-erich-fromm-5b7be96a9fd

Internet Encyclopedia of Philosophy. (n.d.). John Stuart Mill: Ethics. https://iep.utm.edu/mill-eth/

John, F. Kennedy Presidential Library and Museum. (n.d.). Remarks at the University of Kansas, March 18, 1968. https://www.jfklibrary.org/learn/about-jfk/the-kennedy-family/robert-f-kennedy/robert-f-kennedy-speeches/remarks-at-the-university-of-kansas-march-18-1968

Masterson, V. (2022). Degrowth — what's behind the economic theory and why does it matter right now? World Economic Forum. June 15.

Meynen, N. (2023). What is degrowth (and more importantly, what is it not)? META, June 14. https://meta.eeb.org/2023/06/14/what-is-degrowth/

YouTube. (2021). Robert Kennedy on measuring wealth (1968). https://www.youtube.com/watch?v=eXA6BqF2XfA

Circular Economy

I
n today's world, the economy takes all kinds of materials from the earth, makes all kinds of products from them, and then we humans throw them away as garbage. This process is linear; we live in a linear economy. A circular economy (CE) means moving away from our current throwaway economy to one in which products and materials are used for as long as possible. An illustrative example of the current take, make, and throw away culture is the disposal of approximately 1.5 trillion beverage containers per year after the product has been used only once. This huge number represents an enormous amount of material that can be collected, reused, or recycled. Practicing a CE means that valuable materials, as well as energy, water, and petroleum (the basis for plastics), can be saved, thus reducing GHG emissions.

A CE is based on three principles: elimination of waste and pollution, circulation of all products and materials (second-hand, recycling), and regeneration of nature. By applying these three principles, the CE

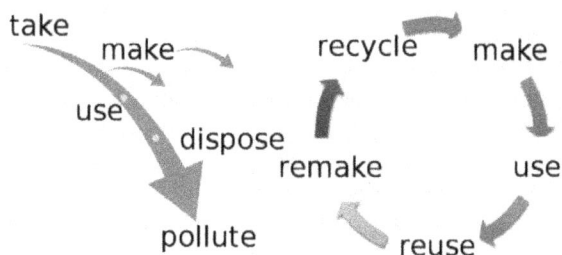

Figure 20.   Linear versus circular economy. A simple diagram to contrast the "take, make, waste" linear approach to the circular economy. (Source: Catherine Weetman, CC BY-SA 4.0)

system decouples the consumption of all finite materials from global economic activity. To follow this resilient path, we must avoid waste as much as possible, bring into the circular flow all the resources provided by our planet, and simultaneously regenerate nature, our great supplier. Sharing wherever possible, repairing, and restoring are other key elements of a CE.

Repairing and reusing materials was common in ancient times and is still common in most so-called underdeveloped countries. A 2021 report on the CE states that only 8.6% of the economy is circular today. The transition to a fully CE will require innovation in system design and constant collaboration and evolution within value chains around the globe. The CE is the cornerstone for an ecologically safe and socially just society in a healthier environment. One example of CE providers is the company Ecovative. The company replaces styrofoam cardboard boxes and packaging with its own mushroom (Mycelium) based products. The company's portfolio also includes 100% vegan, leather-like materials for the fashion market.

Today, 70% of our GHG emissions are associated with the handling of goods, i.e., the extraction, transportation, and processing of raw materials:

- Housing: construction and maintenance of our housing (which creates the largest overall GHG footprint). There is also a great need for action here for effective energy use.
- Communications: everything from data centers to our mobile devices. Expanding connectivity enables a circular economy that eliminates travel and creates better infrastructure.
- Mobility: accounts for a large part of the GHG footprint. This includes the materials used to build mobility equipment (airplanes, cars) and the fossil fuels those vehicles consume.

- Healthcare: Costs are rising as our (Western) population ages and other countries' populations continue to grow. Aging buildings need renovation, and new X-ray and MRI equipment, hospital equipment (beds), pharmaceuticals, disposables, and home care equipment need to be purchased.
- Services: education, public services (government, banking, insurance), office equipment, such as computers, etc.
- Consumer goods: kitchen appliances such as refrigerators and stoves, textiles, clothing, dyes, paints, and chemicals.
- Nutrition: huge footprint (!) — Grains and livestock involve short life cycles but have a big impact (meat consumption).

The Circularity Gap Report finds plenty of reasons to hope that the CE can help us meet the next set of climate goals, even though recent trends show that the numbers for CE adoption are going down rather than up. Hope can be found in the experience almost everyone had during the COVID-19 pandemic. That global crisis showed us what is possible; from governments to citizens, we are now equipped with the knowledge that fundamental change is feasible.

A CE achieves more with less. By applying smart CE strategies in the private, regional, and economic sectors, we can reduce excessive resource consumption and GHG emissions while also living more conscious and secure lives. All we need to do is change our own behavior and view the Earth's resources as a precious gift to be treated with care.

References

Circle Economy Foundation. (2023). The circularity gap report 2023. https://www.circularity-gap.world/2023

ecovative. https://www.ecovative.com/

Ellen Macarthur Foundation. (n.d.). What is a circular economy? https://www.ellenmacarthurfoundation.org/topics/circular-economy-introduction/overview

European Parliament (2023). Circular economy: definition, importance and benefits. May 24. https://www.europarl.europa.eu/news/en/headlines/economy/20151201STO05603/circular-economy-definition-importance-and-benefits

III. Science, Knowledge, and Wisdom

Climate, Science, Ignorance

"The world has changed, and we must change with it."

— Barack Obama,
44th President of the United States (2009–2017)

Climate change is a long-term problem that seems so overwhelmingly large and threatening that it makes people nervous, so they prefer to turn away from addressing it any further. Moreover, for far too many people, it is too complicated and burdensome to ask themselves questions about their behavior in relation to climate change. They instinctively shy away from the problem because they feel that every single aspect of their lives will be affected by climate change. They are afraid of a changing society with different norms and new rules, and especially of the rather gloomy outlook for the future that climate collapse suggests. For them, it is easier to close their eyes and not think about it.

One way to solve this problem of disinterest and fear is to divide the problem into smaller, more manageable segments. These segments are easier to identify, discuss, and then address. The need for transition technologies and other high-impact fixes during the process is obvious. To get everything we need up and running, we have to get everyone on board. All in all, we — meaning each and every one of us — need to move from "business as usual" to "change is usual."

References

Holm, A.O. (2022). Change is the new normal, humans are the only constant. High North News, October 18. https://www.highnorthnews.com/en/change-new-normal-humans-are-only-constant

Sceptical Science. https://skepticalscience.com/

History of Climate Science

M an's influence on the planet is older than many think. According to William Ruddiman, a US paleoclimatologist and geologist, humanity's influence on climate began with the first settlements about 10,000 years ago. However large that initial footprint may have been, it is certain that today's climate crisis was caused by human civilization.

The history of mankind shows that we have gradually acquired all possible means to realize our desire to dominate the world at the expense of nature. The result is that global pollution today threatens all of our planet's vital systems. It is causing devastating damage to all the biological and physical processes that underlie nature and life on Earth. In many parts of the world, pollution has crossed — or is crossing — "planetary boundaries," a general framework that describes limits to the impacts of human activities on the Earth's systems. These tipping points mark irreversible trends. What happens beyond those points — where human-induced changes take the environment out of the more or less stable position that has prevailed for the past 10,000 years — we do not know for sure.

The history of climate science began in the early 19th century. At that time, the existence of ice ages was suspected, and the natural greenhouse effect was discovered. John Tyndall (1820–1893), an Irish physicist, was the first to recognize in 1859 that any change in the amount of water vapor or carbon dioxide (CO_2) in the atmosphere could alter the climate. In honor of Tyndall, the Tyndall Center for Climate Change Research, a university research alliance in England, was established in 2000 "to conduct cutting edge, interdisciplinary research, and provide a conduit between scientists and policymakers."

Figure 21.   Svante Arrhenius (1859–1927). (Public Domain)

In 1896, the Swedish scientist Svante Arrhenius (1859–1927), one of the founders of physical chemistry, calculated that a doubling of atmospheric CO_2 leads to an increase in surface temperature of 5–6°C. The term "greenhouse effect" was first mentioned by John Henry Poynting (1852–1914), an English physicist, in 1909. Because other scientists' experiments could not prove that increasing CO_2 emissions were affecting the climate, the scientific community was not able to confirm the impact of the greenhouse effect or climate change theories for a long time. Mankind was too preoccupied with the two world wars and seemingly unstoppable progress to address something as potentially trivial as the climate, which was constantly changing anyway.

Charles Keeling (1928–2005), an American chemist and geologist, developed the first instrument that could accurately measure CO2 in the atmosphere. On Mount Mauna Loa in Hawaii, he began collecting CO2 data and found that the levels were definitely increasing. The Keeling Curve, a graph showing this trend, was designated as a National Historic Chemical Landmark in 2015 and remains one of the most cited and recognized pieces of evidence for climate change (see Figure 16, page 68).

In 1972, *The Limits to Growth* was published. The report marked a watershed moment in awareness about climate change and rising temperatures around the world. The report's message still holds true today: Even with advanced technologies, the Earth's climate cannot sustain our current economic and population growth.

In 1977, James Black, a senior scientist with the oil company ExxonMobil, reported that the burning of fossil fuels was affecting the global climate, that most aspects of the greenhouse effect needed further research, and that a window of five to ten years was needed to obtain more information on the subject.

The Black Report alarmed the entire oil industry, which then commissioned its own studies on climate change. The conclusion of these so-called studies was that climate change is not man-made but, rather, a natural phenomenon. Thus, for decades, the fossil fuel industry intentionally and successfully raised doubts about climate change. The same thing had happened before in the case of the tobacco industry. The CEOs of the largest United States (US) tobacco companies had publicly and repeatedly denied that smoking was harmful to smokers' health. The tobacco industry persistently spread skepticism about the negative health effects of smoking and encouraged smokers to view their consumption as an individual "right." Similarly, the fossil fuel industry fueled doubt and controversy about climate change, even as the effects of climate change were increasing and almost everyone on Earth could see and even feel those effects.

In 1988, NASA scientist James Hansen (b. 1941) — called the "godfather of climate science" by some — stated at a congressional hearing that the Earth's atmosphere was already warming and that it was 99% certain that this trend was not a natural fluctuation, but was instead caused by an accumulation of CO_2 and other man-made gasses in the atmosphere. Hansen's studies indicated that global warming would not really begin until 2025, with glaciers and polar ice caps melting and sea levels rising dramatically. "It is time to stop waffling so much and say that the evidence is pretty strong that ... global warming has reached a level such that we can ascribe with a high degree of confidence a cause-and-effect relationship between the greenhouse effect and observed warming." In 1988, however, that was not a loud enough wake-up call to stop the self-accelerating spiral toward today's climate emergency.

Also, in 1988, the World Meteorological Organization (WMO) established the Intergovernmental Panel on Climate Change (IPCC). To this day, the IPCC publishes its reports with objective and comprehensive scientific information on human-induced climate change. Until now, IPCC reports have tended to be cautious in their conclusions, but that has changed. The 2021 report is titled "Climate change widespread, rapid, and intensifying" and labels some of the changes, such as sea level rise, as "irreversible over hundreds to thousands of years." We are on the cusp of a new era, the era of climate emergency. When one considers the global impacts of, for example, sea level rise, it becomes clear how far into the climate catastrophe we already are.

References

Brulle, R. (2018). 30 years ago global warming became front-page news — and both Republicans and Democrats took it seriously. The Conversation, June 19. https://theconversation.com/30-years-ago-global-warming-became-front-page-news-and-both-republicans-and-democrats-took-it-seriously-97658

NABU. (2022). News 2022 März Weltklimarat betont Schutz natürlicher Ökosysteme ("News 2022 March Intergovernmental Panel on Climate Change emphasizes protection of natural ecosystems").

The German IPCC Coordination Office. https://www.de-ipcc.de/329.php (in German)

World Meteorological Organization. (2022). March 1. https://www.nabu.de/news/2022/03/31217.html (in German)

YouTube. (2018). James Hansen's 1988 testimony after 30 years. How did he do? https://www.youtube.com/watch?v=UVz67cwmxTM

History of the IPCC

In 1985, the first global scientific conference on climate change was held in Villach, Austria. The conclusions of the scientists present were "very impressive indeed ... and of great importance to mankind." Almost everything we know today, including the conclusions and the actions needed, was published in the conference report — though not in full detail — and it was essentially the first warning about the consequences of human behavior on the planet:

> *The amounts of some trace gasses in the troposphere, notably carbon dioxide (CO2), nitrous oxide (N2O), methane (CH4), ozone (O3), and chloro-fluorocarbons (CFCs) are increasing. These gasses are essentially transparent to incoming short-wave solar radiation, but they absorb and emit long-wave radiation and are thus able to influence the Earth's climate.*

Although the report recommended immediate action by governments and a broad public information campaign on the results of the conference, not much happened. There was another, more pressing problem that needed to be addressed immediately — chlorofluorocarbons (CFCs). As greenhouse gasses (GHGs), CFCs are 14,000 times more potent than CO2. They are better known by the DuPont brand name Freon. Freon had long been used in refrigerators, firefighting, and as a degreasing agent. CFCs destroy the Earth's ozone layer, which is only about 3 millimeters thick — the height of two stacked pennies. The thinning of the ozone layer is a major problem because it increases the amount of ultraviolet radiation that reaches the Earth's surface, thus increasing the risk of skin cancer, cataracts, and other diseases. In response to the dramatic depletion of the ozone layer in the upper atmosphere, especially over polar regions, CFCs were banned in 1987 by the Montreal Protocol. The protocol was ratified by 197 parties (196 countries plus the European Union (EU)), making it the first universally ratified United Nations (UN) treaty.

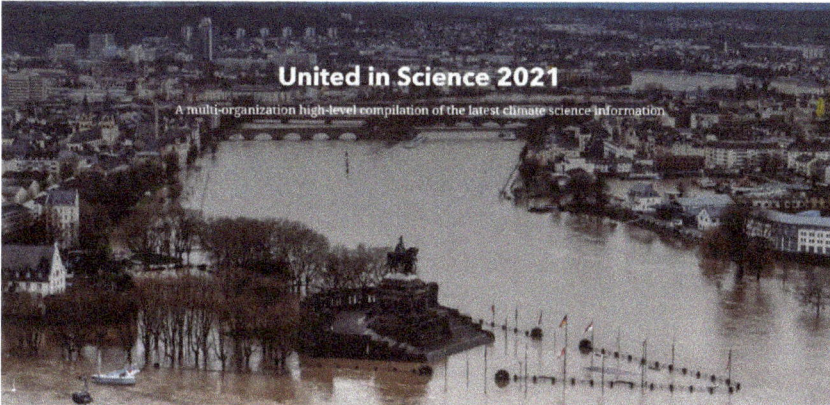

Figure 22.   WMO-Report 2021, United in Science (cover).

After another meeting of more than 300 scientists in Toronto in 1988, the United Nations Environment Programme (UNEP) and the WMO eventually formed the IPCC.

The first IPCC report was published in 1990. In addition to the fact (already published in the 1985 Villach Report) that GHGs in the atmosphere were increasing, the first IPCC report warned that "long-lived gasses require immediate reductions of over 60% … in emissions from human activities" and that "continued emissions of these gasses at present rates would commit us to increased concentrations for centuries ahead." The report projected a rise in the mean global sea level of about 20 centimeters by 2030 and 65 centimeters by the end of the century: "There will be significant regional variations."

However, even this message did not have much impact on national or international policy. After the publication of some supplementary reports in

1992, the second IPCC report (1995), subtitled "The Science of Climate Change," confirmed all the other earlier conclusions. However, it issued a more serious warning: "Projections of future global mean temperature change and sea level rise confirm the potential for human activities to alter the Earth's climate to an extent unprecedented in human history; and the long time-scales governing both the accumulation of greenhouse gasses in the atmosphere and the response of the climate system to those accumulations, means that many important aspects of climate change are effectively irreversible."

In addition to these warnings, the second report recognized its own role as the primary source of scientific and technical information for the UN Framework Convention on Climate Change (FCCC). The real goal of this report was to provide objective information on which to base global climate policy. It was

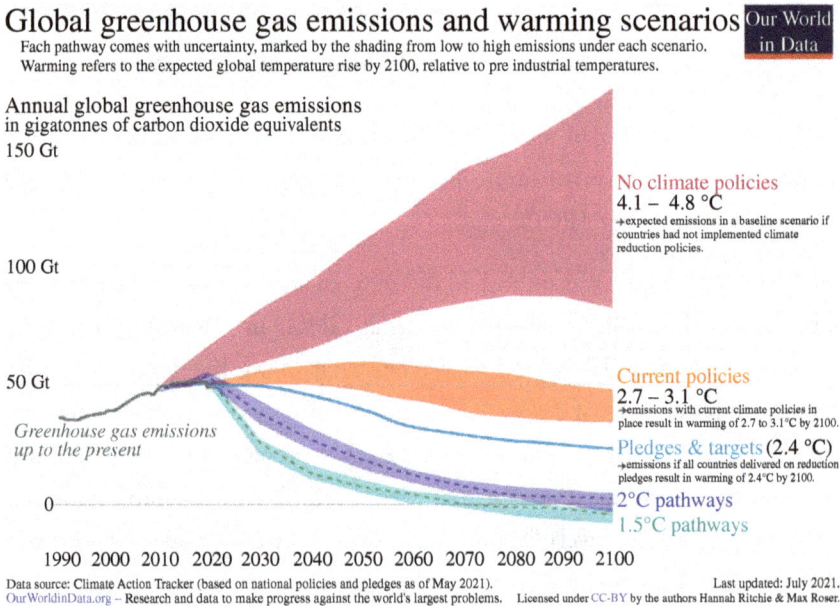

Global greenhouse gas emissions and warming scenarios

Each pathway comes with uncertainty, marked by the shading from low to high emissions under each scenario.
Warming refers to the expected global temperature rise by 2100, relative to pre industrial temperatures.

Our World in Data

Annual global greenhouse gas emissions
in gigatonnes of carbon dioxide equivalents

150 Gt

No climate policies
4.1 – 4.8 °C
→expected emissions in a baseline scenario if
countries had not implemented climate
reduction policies.

100 Gt

50 Gt

Current policies
2.7 – 3.1 °C
→emissions with current climate policies in
place result in warming of 2.7 to 3.1°C by 2100.

*Greenhouse gas emissions
up to the present*

Pledges & targets **(2.4 °C)**
→emissions if all countries delivered on reduction
pledges result in warming of 2.4°C by 2100.

0

2°C pathways
1.5°C pathways

1990 2000 2010 2020 2030 2040 2050 2060 2070 2080 2090 2100

Data source: Climate Action Tracker (based on national policies and pledges as of May 2021).   Last updated: July 2021.
OurWorldinData.org – Research and data to make progress against the world's largest problems.   Licensed under CC-BY by the authors Hannah Ritchie & Max Roser.

Figure 23.   Greenhouse gas emissions and warming scenarios. (Source: Hannah Ritchie, Max Roser, and Pablo Rosado)

imperative that policymakers achieve the FCCC's ultimate goal, articulated in Article 2 of the Convention, of stabilizing GHGs at a level that would prevent "dangerous anthropogenic interference with the climate system."

In 1997, the Kyoto Protocol established the goals of the UNFCCC (United Nations Framework Convention on Climate Change), namely, to reduce GHGs in the atmosphere. It is an international treaty to which the 192 signatory countries have committed themselves. Developing countries, including China, India, and Brazil, were exempted from the treaty, as they were not main GHG contributors during the industrialization period. The Kyoto Protocol recognizes that different countries have different capabilities to combat climate change based on their economic development. It transfers the obligation to reduce emissions to more developed countries. The Protocol binds these developed countries and places a greater burden on them under the principle of "common but differentiated responsibility and respective capabilities." Important additional market-based mechanisms under the Kyoto Protocol include emissions trading, which allows countries to sell excess capacity to countries that have already exceeded their targets.

In 2001, the IPCC's Third Assessment Report (TAR), subtitled "The Scientific Basis," was published. It was much more comprehensive than the two previous reports and contained even clearer conclusions:

- The 1990s has been the warmest decade, and 1998 the warmest year on record since 1861.
- The temperature rise in the 20th century was almost certainly the largest in the past 1,000 years.
- Human-induced climate change will certainly continue for many centuries to come.

Throughout the scientific world, the report was hailed as "truly amazing," a "standard reference," and "impressive and respectable." The IPCC "has conducted arguably the largest, most comprehensive and transparent study ever undertaken by humankind. The result is a work of substance and authority that only the stupid make fun of."

The fourth IPCC report was published in 2007. It was more than a thousand pages long and more unequivocal than any previous report: "Warming of the climate system is unequivocal, as is now evident from observations of increases in global average air and ocean temperatures, widespread melting of snow and ice, and rising global average sea level." As for where the observed sea level rise is coming from, the report clearly stated that "losses from the ice sheets of Greenland and Antarctica have very likely contributed to sea level rise over 1993 to 2003."

The impacts of climate change on the world have been described as follows: "At continental, regional, and ocean basin scales, numerous long-term changes in climate have been observed. These include changes in arctic temperatures and ice, widespread changes in precipitation amounts, ocean salinity, wind patterns, and aspects of extreme weather including droughts, heavy precipitation, heat waves, and the intensity of tropical cyclones."

The fourth IPCC report issued an even more urgent warning about the impacts of climate change. "Unmitigated climate change would, in the long term, be likely to exceed the capacity of natural, managed and human systems to adapt." This was a clear alarm signal, but it did not have the intended effect. In addition, the report contained some minor errors. One paragraph about the projected timing of Himalayan glacier melt was incorrect. This led to calls for a review of the entire process used to formulate the report.

James E. Hansen, the "godfather of modern climate science," also criticized the report, saying, "The IPCC notes that they are unable to evaluate possible dynamical responses of the ice sheets, and thus do not include any possible 'rapid dynamical changes in ice flow'. Yet the provision of such specific numbers for sea level rise encourages a predictable public response that the projected sea level change is moderate, and smaller than in IPCC (2001). Indeed, there have been numerous media reports of 'reduced' sea level rise predictions, and commentators have denigrated suggestions that business-as-usual greenhouse gas emissions may cause a sea level rise of the order of meters. [...] The IPCC is doing a commendable job, but we do need something more."

Richard Tol (b. 1969), a former IPCC author, admonished the entire process of IPCC reports. In particular, Tol criticized the process of having report summaries — written specifically and explicitly for policymakers — agreed upon between scientists and government officials: "Academic quality should be the guiding principle in selecting authors. As a check, the committees that nominate and select authors should publish their proceedings. The review editors should become more independent and gain the right to reject chapters that are not properly revised. The alternative is a gradual erosion of the quality, prestige and, eventually, influence of the IPCC."

In 2013, the fifth IPCC Report was published. It presented "clear and robust conclusions in a global assessment of climate change science — not the least of which is that the science now shows with 95 percent certainty that human activity is the dominant cause of observed warming since the mid-20th century." The report confirms that warming in the climate system is unequivocal, with many of the observed changes being unprecedented over decades to millennia. In more than 1,500 pages, the report provides evidence of warming in the climate system, increases in energy stored in the climate system, rises in mean global sea level, ice loss, and increases in atmospheric concentrations of carbon dioxide, methane, and nitrous oxide (laughing gas). The main message of this report is unequivocal; namely, that mitigating "climate change will require substantial and sustained reductions in greenhouse gas emissions."

In 2015, the Paris Agreement — a legally binding international treaty on climate change — was adopted by 196 parties. The agreement calls for huge economic and social change across the globe. Its goal is to limit global warming to well below 2°C and to continue efforts to limit temperature increases to 1.5°C. Signatory states are committed to reducing their GHG emissions as rapidly as possible and "consistent with the best available science" to achieve climate neutrality by 2050 ("Net Zero"). Each signatory country must submit its Nationally Determined Contributions (NDCs) on a five-year cycle with increasingly ambitious climate targets. These reports must be accompanied by information on how to reduce emissions, increase carbon sinks, and manage the unavoidable impacts of climate change.

In 2021, the IPCC Sixth Assessment Report was published. It runs to almost 4,000 pages, and the key finding is crystal clear — climate change is already affecting all inhabited regions of the world, with human influence contributing to many observable changes in weather and climate extremes. According to this report, the future of our climate looks pretty bleak:

- Global warming of 1.5°C or 2°C will be exceeded in the 21st century unless there is a profound reduction in CO2 and other GHG emissions in the coming decades.
- Many changes in the climate system are becoming larger in direct relation to increasing global warming — heat extremes, marine hot tides, heavy precipitation, droughts, intense tropical cyclones, the decline of Arctic sea ice, snow coverage, and permafrost.
- The report also notes that a 2°C increase is much worse than "only" a 1.5°C increase and that all regions of the world will be affected.

References

European Council and Council of the European Union. (n.d.). https://www.consilium.europa.eu/de/policies/climate-change/paris-agreement/

Hansen, J.E. (2007). Scientific reticence and sea level rise. *Environmental Research Letters*, **2**:024002. https://iopscience.iop.org/article/10.1088/1748-9326/2/2/024002/pdf

Houghton, J.T., Ding, Y, Griggs, D.J., *et al.* (2001). A report of Working Group I of the Intergovernmental Panel on Climate Change: Summary for policymakers. Intergovernmental Panel on Climate Change. https://www.ipcc.ch/site/assets/uploads/2018/03/WGI_TAR_full_report.pdf

Houghton, J.T., Meira Filho, L.G., Callander, B.A., *et al.* (eds.) (1996). Climate change 1995: The science of climate change. Intergovernmental Panel on Climate Change. https://www.ipcc.ch/site/assets/uploads/2018/02/ipcc_sar_wg_I_full_report.pdf

Intergovernmental Panel on Climate Change. (2023). Sixth Assessment Report. https://www.ipcc.ch/assessment-report/ar6/

Tol., R.S.J. (2007). Biased policy advice from the Intergovernmental Panel on Climate Change. *Energy and Environment*, **18**(7–8). https://web.archive.org/web/20101009033936/http://www.fnu.zmaw.de/fileadmin/fnu-files/publication/tol/RM7422.pdf

UNEP/WMO/ICSU. (1985). Statement by the UNEP/WMO/ICSU International Conference on the assessment of the role of carbon dioxide and of other greenhouse gases in climate variations and associated impacts villach, Austria, 9–15 October. http://www.climatechange.lk/DNA/data/Statement%20by%20the%20UNEP%20on%20Villach_1985.pdf

United Nations Climate Change. (n.d.). Climate change 2001: The scientific basis. https://www.ipcc.ch/site/assets/uploads/2018/03/WGI_TAR_full_report.pdf

United Nations Climate Change. (n.d.). The Paris Agreement. https://unfccc.int/process-and-meetings/the-paris-agreement

United Nations Climate Change. (n.d.). What is the Kyoto Protocol? https://unfccc.int/kyoto_protocol

United Nations Climate Change. (n.d.). What is the United Nations Framework Convention on Climate Change? https://unfccc.int/process-and-meetings/what-is-the-united-nations-framework-convention-on-climate-change

Conference of the Parties

The Conference of the Parties (COP) is the supreme decision-making body of the United Nations Framework Convention on Climate Change (UNFCC). This convention came into force on March 21, 1994. The goal of the Convention is to prevent "dangerous" human interference with the climate system. The COP reviews national communications and emission inventories submitted by Parties. The COP then assesses the impact of the measures taken. COP meetings are held every year, the first one being in Berlin in 1995. Typically, the COP meets at different locations, decided by the Bureau of the COP. A Party may also offer to host the next meeting, as was the case for COP26, which was held in Glasgow in 2021. COP27 took place in Egypt in 2022, and COP28 (2023) in Dubai.

In recent decades, climate change has been the subject of a number of books, plays, and films. *An Inconvenient Truth* is a 2006 documentary by Davis Guggenheim about former US Vice President Al Gore's campaign to raise awareness about climate change. It is based on a slide show lecture by Al Gore that he had given over a thousand times. The film was very successful, grossing over US$50 million and winning two Oscars. Al Gore, the main character of the film, emphasizes that global warming is not only a political issue but also a moral one. He closes the film with some remarkable words that show us who the real culprits of climate change are — namely, us — and how we could, if we wanted to, end the deadly spiral by making "choices to bring our individual carbon emissions to zero": "Each one of us is a cause of global warming, but each one of us can make choices to change that with the things we buy, the electricity we use, the cars we drive; we can make choices to bring our individual carbon emissions to zero. The solutions are in our hands, we just have to have the determination to make it happen. We have everything that we need to reduce carbon emissions, everything but political will."

References

United Nations Climate Change. (n.d.). Conference of the Parties (COP). https://unfccc.int/process/bodies/supreme-bodies/conference-of-the-parties-cop

United Nations Climate Change. (n.d.). What is the United Nations Framework Convention on Climate Change? https://unfccc.int/process-and-meetings/what-is-the-united-nations-framework-convention-on-climate-change

United Nations Framework Convention on Climate Change. (2008). Kyoto Protocol. Reference manual: On accounting of emissions and assigned amount. https://unfccc.int/sites/default/files/08_unfccc_kp_ref_manual.pdf

World Meteorological Organization. (2021). Climate change and impacts accelerate. September 16. https://public.wmo.int/en/media/press-release/climate-change-and-impacts-accelerate

Climate Scientists

Today, global warming is accelerating at an unprecedented rate. The ice caps of the Earth's poles, the Greenland ice sheet, and the ice of the glaciers are melting at an alarming rate. The ice is also melting much faster than most scientists or anyone else thought possible.

Now, to immediately figure out how to address an unprecedented global emergency, an unprecedented joint venture could potentially find answers to these challenges. There are already model science centers, such as the European CERN in Geneva, where more than 3,000 scientists have spent decades searching for the next previously unknown phenomenon in the subatomic world. Why shouldn't it be possible to create such a joint world science center for climate change? What is lacking? Does New York City have to be submerged before we get serious about finding solutions?

Sir David Attenborough (b. 1926) is the world's best-known scientist and, for decades, has been a tireless campaigner for biodiversity, climate change mitigation, and reduced meat consumption. He is the most charming spokesman and enthusiast of the global movement to combat the destruction of the planet's splendor and wealth.

Michael E. Mann (b. 1965), an American climatologist and geophysicist, is Distinguished Professor of Atmospheric Science at Pennsylvania State University in the US. In 1999, Mann and his team produced the well-known "hockey stick graph," mapping the rapid warming of the Earth in the 20th century. In the years that followed, Mann's results were vigorously opposed, especially by the oil industry. But since 2007, when the fourth IPCC report was released, it has become common knowledge that temperatures in the last hundred years have been the highest in the past

1,300 years and that CO2 levels in the atmosphere are now rising exponentially. Mann is one of the highest-profile campaigners for rapidly shifting toward a fossil-fuel-free future. In his recent book, *The New Climate War*, he discusses the actions of the fossil fuel industry to delay relevant action on climate change. He is also confident that everything we need in the "fight to take back our planet" is already in place and that a "clean energy revolution and climate stabilization are achievable with current technology. All we require are policies to incentivize the needed shift." Once the strategies and measures have been implemented, "the tide may finally be turning in a hopeful direction."

Stefan Rahmstorf (b. 1960) is deputy head of the Earth System Analysis Research Department at the renowned Potsdam Institute for Climate Impact Research (PIK) in Germany. Trained as a physicist and oceanographer, his research focuses on ocean circulation and Earth system modeling. He is a busy, versatile, and well-known science communicator and, as of 2020, the climate scientist with the most Twitter (now known as "X") followers in Europe. A major area of his research is monitoring the Atlantic Meridional Overturning Circulation (AMOC). This is a massive thermal conveyor belt in the deep ocean that transports more than 20 times as much water as all of Earth's rivers combined. Part of this system is the Gulf Stream, which influences the climate in various places around the world — the weather in Europe, the sea level on the east coast of the US, and the intensity of tropical cyclones in the Atlantic.

Rahmstorf and his team have discovered that the Gulf Stream system has slowed down since the 1950s. A weakening of this gigantic heat pump would have dramatic consequences for the population in Europe and beyond. Entire marine ecosystems would be threatened with collapse; if the Gulf Stream's circulation pump were to sputter, not enough nutrients would be flushed from the depths to the surface. Even a partial weakening would potentially lead to a collapse of plankton stocks, with consequently catastrophic effects on fish populations in the Atlantic. Rahmstorf stated: "When the ice sheet in Greenland melts, freshwater flows into the North Atlantic. This lowers the salinity at the surface, and thus the density, so that the cold surface water sinks more slowly into the depths. This slows down the circulation. And global warming means more and more freshwater is entering the water cycle." Rahmstorf also warns that there is a widespread loss of control over the further course of the climate crisis because global warming is likely to release methane gas from the oceans and permafrost regions, effects that can no longer be controlled or limited. These are concrete scientific findings that point to the very real danger that many irreversible tipping points will soon be crossed.

Johan Rockström (b. 1965) is director of PIK. Together with Sir David Attenborough, Rockström appeared in the Netflix documentary *Breaking Boundaries* in 2021. The film paints an undeniably bleak picture of our future. We must act quickly to prevent an apocalyptic scenario like the one in the classic film *Mad Max*. *Breaking Boundaries* is mainly about so-called tipping points. Today, we know that these events may be more likely than previously assumed by politicians and even climate scientists. Tipping points have major impacts and are interconnected across different biophysical systems. If they are crossed, it can and will lead to long-term, irreversible changes in the world.

Tragically, we have all known for a long time what we must and must not do to keep the earth both healthy and stable. We need to generate energy exclusively from renewable sources everywhere and in a decentralized manner, and we must stop using fossil fuels. We need to preserve forests and re-water the world's wetlands, especially peatlands. We need to plant lots of trees to have forests that are established and healthy in a few decades because they are good and cheap carbon reservoirs. We are not doing this for ourselves but for our children and grandchildren.

Klaus Hasselmann (b. 1931) is the 2021 Nobel Laureate in Physics. He is the "visionary of climate science." Hasselmann has contributed significantly to our understanding of climate catastrophes by analyzing the chaos that we used to call weather. While working at the Max Planck Institute for Meteorology in Hamburg in the 1970s, Hasselmann showed that climate models could be reliable even in extremely changeable weather. He developed a model that could convert seemingly unpredictable weather fluctuations into reliable forecasts. Hasselmann's achievements may be illustrated in this way: A dog on a leash always remains in a relationship to the person walking with it. In the same way, the weather always remains in relation to the climate. In 1988, Hasselmann warned that in the next 30 to 100 years, depending on how much fossil fuel we consume, we would face very significant climate change — "Climate zones will shift, and precipitation will be distributed differently. We should realize that we are entering a situation where there is no turning back."

The other two winners of the 2021 Nobel Prize in Physics are Syukuro Manabe, a meteorologist at Princeton University, and Giorgio Parisi, an Italian theoretical physicist. Parisi received the prize for his discoveries about "how apparently random phenomena are governed by hidden rules." All three scientists received the Nobel Prize for their work in weather and climate modeling, as well as the impact of humans on global warming. They laid the foundation for our understanding of the role of human activities and GHGs in climate change.

Fridays for Future — It seems that scientists like David Attenborough have found their "natural partner" in the younger generation, with the iconic and emblematic Greta Thunberg (b. 2003) as the main character, who said: "Fossil fuel companies have been, for decades longer than I have been alive, the largest contributors to the climate crisis that affects my generation today — all in pursuit of profits and growth." In *The New Climate War*, Michael Mann holds them to account and shows us how we can take the bold steps we must all take together to win the battle to save this planet." The young generation has pushed the issue of climate change into the mainstream media, and with very good reason — their lives are at stake. Either we all find ways to realize a "greener" world, or, if we continue on our current track, we may all witness the destruction of the planet's biosphere unfolding before our eyes, the prelude of which we are already witnessing every day.

The Fridays for Future movement and many other organizations, such as Extinction Rebellion, ATTAC, and dozens of other nongovernmental organization (NGOs), have already mobilized millions of mostly young people around the globe. The ongoing and ever-growing community of these organizations is an undeniable expression of the awareness of this younger generation and how much they care about a sustainable environment. Luisa Neubauer (b. 1996) is the German face of Fridays for Future and has proposed new goals for a new society: "Prosperity and well-being is different from endless consumption that destroys our own livelihoods. We need change."

So, how do we need to change?

The Centre for the Understanding of Sustainable Prosperity (CUSP), a "cutting-edge" research organization, is led by Professor Tim Jackson (University of Surrey). The organization's vision is that people everywhere should have the opportunity to flourish as human beings, where prosperity means providing society with a credible and inclusive vision of social progress.

Lorraine Whitmarsh, a CUSP partner, is an environmental psychologist and director of the UK Centre for Climate Change & Social Transformations (CAST) at the University of Bath. Whitmarsh was awarded the MBE (Member of the Order of the British Empire) in the Queen's New Year Honors in 2021 in recognition of her work on behavior change and public engagement for a more sustainable future. Her research projects include studies on energy-efficiency behaviors, behavior change, waste reduction, low-carbon lifestyles, and responses to climate change. Her research focuses on action and innovation for a more sustainable future to address the impacts of climate change.

Richard Thaler (b. 1945), an American economist and winner of the 2017 Nobel Prize in Economics, has a simple rule: "My mantra is if you want to get people to do something, make it easy. Remove the obstacles." So how, for example, can consumption habits be changed to establish a healthier, meat-free diet?

Consumption habits take a lot of getting used to; bringing about a change in behavior is complicated. Availability, taste, and price tend to have a greater impact on human behavior than their decision's potential impact on the environment. The "nudge" approach to behavior change advocates contextual changes to make new habits easier to achieve with a "nudge." Example: In an experiment in Scandinavia, low-carbon foods were placed at the top of the menu to raise awareness. In Switzerland, another field experiment sought to make renewable energy contracts the default option. However, such "kick-off" strategies also carry a risk. If they are introduced at a very high level, such as energy spending, they could attract the attention of those who are already opposed to any regulatory incentives. Appealing to emotions is also a very useful way to motivate behavior change "because they are immediate and directly felt." Overall, these experiments have already determined which behavior change strategies are most likely to be successful.

The business sector has found that consumers are more likely to choose plant-based dishes after the introduction of a wider range of plant-based dishes or even when the "placement" of such dishes on a menu is changed and improved. Placing vegetarian or vegan dishes on par with meat dishes encourages people to try new things, as customers see the dishes as equal to traditional meat dishes. With millions of people calling themselves flexitarians, the opportunity to sell plant-based dishes is huge. This demonstrates one of the many ways — pricing is another — to bring plant-based, climate-smart foods to the marketplace.

In a 2021 food eco-labeling pilot conducted by a team at Oxford University, each dish was labeled with the letters A through E to indicate which dishes had a higher (E) or lower (A) environmental impact. These ecolabels had a statistically significant impact on reducing the overall environmental impact of the average purchase. If these labels became more widespread, vendors and suppliers could become more aware of this issue and change their offerings to more environmentally friendly products. Finally, there is another important rule: More consumption means more harm, so less consumption is much more important than product selection.

References

Centre for the Understanding of Sustainable Prosperity. https://cusp.ac.uk/

Foundation Earth. https://www.foundation-earth.org/about-us/

Intergovernmental Panel on Climate Change. https://www.ipcc.ch/

Max Planck Institute for Meteorology. (2021). Physics Nobel Prize 2021 for Klaus Hasselmann. October 6. https://www.mpg.de/nobel-prize-physics-2021/klaus-hasselmann

People's Agreement of Cochabamba. https://www.climateemergencyinstitute.com/uploads/Peoples_climate_agreement.pdf

Rahmstorf, S. (2008). The idea that humans can change and are in fact changing the climate of our planet has developed gradually over more than a hundred years. Potsdam Institute for Climate Impact Research, January. https://www.pik-potsdam.de/~stefan/Publications/Book_chapters/Rahmstorf_Zedillo_2008.pdf

Roebeling, R., John, V., Prieto, J., et al. (2021). Observing the cooling of the North Atlantic Ocean during the last decade, using weather satellites. EUMETSAT, September 6. https://www.eumetsat.int/melting-greenland-ice-sheet-cools-north-atlantic-ocean

University of Bath. Lorraine Whitmarsh, FHEA MBE. https://researchportal.bath.ac.uk/en/persons/lorraine-whitmarsh

World Meteorological Organization. https://public.wmo.int/en/about-us

Reason and Responsibility

It is necessary to create sober, patient people who do not despair in the face of the worst horrors and who do not become exuberant with every silliness. [...] Pessimism of the intellect, optimism of the will.

— Antonio Gramsci (1891–1937), Marxist philosopher

Curiosity and science are human responses to nature. We are curious about what is real, where it comes from, and what it is made of. We are also curious about ourselves, sometimes, in moments of calm in this busy, hectic world. Then we realize that we are lucky to exist at all. Our life is the most remarkable thing we have, and it is a remarkable piece of luck that we can live our lives. We know of no other place in the universe where intelligent life exists. On Earth, it took four billion years for our civilization to emerge, which could mean that intelligent life is very, very rare. We are now at a point in our history where we must ask ourselves: What are the priorities of our civilization in the face of a climate catastrophe? Is it about our own well-being (which is certainly part of the answer), or are there larger, more complex needs in our world?

One of the many problems facing our civilization today is that we do not realize how valuable we really are. In our galaxy, there are approximately 100–400 billion stars. There are old and dark stars that are difficult to detect, orbited by trillions of planets that could potentially harbor life. And yet, it is also possible that the number of civilizations that exist out there could be counted on one hand. It is even possible that there is only one, and that is us. We are the ones asking the questions, "Are we alone in the universe?" or "What is the point of it all?" At this point, reason meets responsibility. Will we allow or participate in the destruction of reason in the galaxy?

Figure 24.   Are we alone? (Source: NASA, Tracy Caldwell Dyson, 2010)

Stupidity and Ignorance

Two things are infinite: The universe and human stupidity; and I'm not sure about the former.

— Albert Einstein (1879–1955), physicist [attributed to]

In an interview on Danish television, a bright eight-year-old boy said that, in his opinion, people who do not recognize obvious signs of climate change as such should be called "stupid." What did he really mean when he said "stupid" — silly, daft, or foolish?

Stupidity is a lack of intellectual processing ability, as measured by IQ tests. It is a distinction of gross cognitive incapacity. The German philosopher Immanuel Kant (1724–1804) made a sharp distinction between stupidity and simple-mindedness: "The simpleton is he who cannot grasp much through his understanding; but he is not therefore stupid." A stupid person has only weak cognitive abilities, he is "dull in the head" and has a "weak mind" (Robert Musil).

In Latin, "stupidus" refers to a person in a state of stupor, astonishment, or confusion. According to Kant, stupidity is "the absence of judgment" or an insufficiency of understanding in the absence of judgment. In the same sentence, Kant said about stupidity that "for such a failing we know no remedy." Kant was certain that even very learned people with highly developed minds can suffer from this disease, any "physician, judge, or statesman ... [with] many fine pathological, juridical, or political rules in his head." Stupid people may know all the universal rules and all the concepts and regulations in their own professions and disciplines, and yet they may be incapable of applying their knowledge to other circumstances.

Foolishness is something else again. It can be attributed to someone who falls for telephone scams and similar tricks but may not be stupid at all. Stupidity can be understood as a failure of prudence, and foolishness as a failure of intelligence. One could also follow Kant's definition of stupidity as a "failure of judgment," or even more precisely and properly adapted to our modern world of media and conspiracy theories, as the inability or unwillingness to make judgments.

Let us take a look at other scholars: Erasmus of Rotterdam (Desiderius Erasmus, 1469–1533), the greatest scholar of the Dutch Renaissance, wrote: "The most clumsy stupidity, the most perverse perversity, always finds the most admirers and lovers on earth, [...] because almost all men pay homage to folly. Ignorance has two great advantages: first, it is perfectly compatible with self-love, and second, it enjoys the admiration of the crowd."

Artists have always been concerned with human stupidity. Gustave Flaubert's *Bouvard and Pécuchet* (published in 1881) is a wonderful example, and the Austrian writer Robert Musil (1880–1942) is another of the many artists who have written about stupidity. In 1937, when the Nazis were already at the door to invade Austria, Musil delivered the remarkable speech "On Stupidity." In these lines, there is an interesting association of stupidity and vanity: "A certain intellectual-mental petty bourgeoisie is completely shameless about the need for arrogance (vanity) as soon as it acts under the protection of a party, nation, sect or art movement and one is allowed to say 'we' instead of 'I.'" Musil added, with a clear wink at the Nazis, that "there is a great tendency in the world for people, when they come together in large numbers, to allow themselves to do everything that is forbidden to them as individuals." Applying these words to today's world, one can discover certain similarities. For example, on Telegram, an Internet platform, almost anything is allowed, even if it is very, very stupid or outright wrong and defamatory.

The Basic Laws of Human Stupidity is a remarkable book published in 1976 by Italian writer Carlo Cipolla (1922–2000). In his book, the author tries to explain how stupid people function and why we should beware of them: "The problem with stupid people is when they take up positions of power. And they occupy all positions in society, from top to bottom. In a previous era, the influences on social organisation came from the church, class, and caste. Now we have political parties and bureaucracy, and lieu of religion we have democracy. Within a democratic system, general elections are a most effective instrument to ensure the steady maintenance of the fraction [of stupid people] among the powerful. Hence a stupid person with power can be most threatening." Cipolla has formulated five basic laws of stupidity:

1. Always and inevitably, everyone underestimates the number of stupid individuals in circulation.
2. The probability that a person is stupid is independent of any other characteristic of that person. Education, wealth, or status have nothing to do with it.
3. A stupid person is someone who causes losses to other people while himself deriving no gain.
4. Non-stupid people always underestimate the damaging power of stupid people and forget that at all times and under all circumstances, dealing with stupid people is a very costly mistake.
5. A stupid person is the most dangerous type of person.

"Are we living in the age of stupid?" An article in *The Guardian*, a British newspaper, began with this question. In it, the authors quote American writer Charles Bukowski: "The problem with the world is that intelligent people are full of doubts, while the stupid ones are full of confidence." So how is it that despite having more information at our fingertips today than in the whole of human history, so many people are still easily misled by misinformation?

"Stupid people" ignore reason. They are easily misled; they fall for gossip and are seduced by rumors and conspiracy theories. Unfortunately, such people are everywhere. For example, in the context of the COVID-19 pandemic, everyday vaccine deniers had their own methods to defend their beliefs. They simply pointed the finger in another direction, claiming: "These idiots are all so gullible. They're all getting vaccinated! Horror!" Many stupid people have had a good education and are convinced of their own intelligence. However, even intelligent people can be stupid sometimes. And, of course, many people are sometimes a bit clueless or simply wrong or insufficiently informed.

"Stupid people are unaware of their stupidity." The problem underlying this statement is that our technical and constructive abilities are far ahead of our social and intellectual abilities. The rapidly evolving global economic system requires technological minds, not smart people asking questions. But right now, in the face of mounting crises, the situation demands that we fight our collective stupidity in order to develop collective intelligence. First, we must rid ourselves of the biggest lie of all: that continuous growth is possible in a finite world. We have reached the limits of growth and must escape the toxic spiral of growth, consumption, and greed. This "chug and chuck" attitude has ruined our collective psyche and caused enormous damage to nature.

Of course, we also have to confront all the other lies on which our society's collective stupidity (or ignorance) is based: that only simple political solutions like "America first" or the Russian "Z" symbol are correct; that only large corporations can survive, while small, community-oriented businesses must die out; and that the existing global tax and monetary system is truly sustainable.

People who question the system are met with great hostility. Ironically, abstract thinking about politics, history, and the functioning of society is based on appreciation, caring, openness, or even sympathy for other points of view. The real problem is that our society produces people who are lonely, hostile, and defensive. This leads to selfish, stupid, and dogmatic thinking. Particularly troubling is the growing phenomenon of anti-factual thinking. Even when they hear convincing arguments against the conspiracy theory in which they believe, "true believers" are not impressed by any argument. Such people will not let anything dissuade them from their convictions because they do not have a self-critical mind guided by logic and rationality. Einsteins are rare; stupid people are everywhere.

We are often reluctant to change our minds about certain issues, especially if the views we hold have become part of our self-image or way of life. Changing would mean admitting that we were wrong, and that admission could be painful to our fragile egos. People's beliefs in this regard are often just prejudices or products of certain socialization, indoctrination, or peer pressure.

In reality, most people are not really interested in arguments. They simply have other interests, which mostly revolve around personal questions ("How will I get a better job?") and/or leisure activities ("Where will we go on vacation?"), while others focus their activities on social media. Most people will not fight for their own economic interests. The worsening economic, political, and climatic situation around the world, with increasing numbers of displaced populations, leads some people to blame immigrants for the situation.

The infrastructure of social media makes it very easy for people to surround themselves exclusively with like-minded people and like-minded news sources. Together with a kind of intellectual laziness and lack of curiosity about the beliefs of others, the self-fixated personality type generated by neoliberal social structures will prevail. The real problem here lies in the power of the media world, which is partly controlled by right-wing media moguls (e.g., Rupert Murdoch) or by corporations interested only in shareholder value and high profits.

Human beings have a remarkable capacity for self-deception. Our minds are profoundly opaque to themselves; in many ways, we are constantly deceiving ourselves without knowing it. We constantly tell ourselves stories to bolster our self-esteem. We pass over events that might contradict our self-interpretation and refuse to deal with unpleasant things. We are afraid of change because it would challenge our deeply held beliefs about society. Changing old beliefs would overturn the image of life that people want to retain in their minds. It is sheer fear that prevents us from taking a different path into unknown territory, even if that path is the only reasonable one.

So what did the eight-year-old boy mean when he called climate-change deniers "stupid"? Perhaps he was referring to something like a "blind spot" in people's minds — something they do not think or will not talk about. This could be called the "Don't Look Up" syndrome. They are deeply afraid to look behind this blind spot. It is the fear of the reality of the elephant in the room. Let us face the music, be strong, and start talking about the elephant — the climate crisis!

References

Buckmaster, L. (2021). Idiocracy: A disturbingly prophetic look at the future of America — and our era of stupidity. *The Guardian*, July 18. https://www.theguardian.com/culture/2021/jul/19/idiocracy-a-disturbingly-prophetic-look-at-the-future-of-america-and-our-era-of-stupidity

Ludwig, P. (n.d.). Objectivity's blind-spot: The Dunning-Kruger effect. Procrastination.com. https://procrastination.com/blog/31/objectivity-dunning-kruger-effect

Pagel, M. (n.d.). Breaking the wall of collective stupidity. Falling Walls. https://falling-walls.com/discover/videos/breaking-the-wall-of-collective-stupidity/

Tettamanzi, A.G.B. and da Costa Pereira, C. (2014). Testing Carlo Cipolla's laws of human stupidity with agent-based modeling. IEEE/WIC/ACM International Conference on Intelligent Agent Technology, IAT 2014, Warsaw, Poland. pp. 246–253.

Wikipedia. (n.d.). Dunning–Kruger effect. https://en.wikipedia.org/wiki/Dunning%E2%80%93Kruger_effect

Knowledge and Wisdom

The only true wisdom is in knowing you know nothing.

— Socrates (469–399 BC), Greek philosopher

K nowledge is what we know; it grows and changes as we get older. Intelligence is the basis and measure of what and how much we understand. Intelligent people know how to analyze and interpret things they encounter, how to develop concepts, and how to abstract from reality. Both knowledge and intelligence can be measured to some degree. The same is not true of wisdom. Wisdom is the quality of insight one has, and it operates in a completely separate realm. One can be intelligent without being wise. Simply put, wisdom could be defined as the ability to understand the nature of man and the world as a whole.

Today, there is an alarming number of people around the world who have crazy or even dangerous ideas. They reject the consensus on human-induced climate change as a hoax and act as though vaccinations are toxic to them. It is obvious that a large number of people are unable to act rationally. There is also a growing number of conspiracy-theory believers in the world ("QAnon"), and some of them are smart, educated, and influential. Paul Krugman, winner of the 2008 Nobel Prize in Economics, called them "zombies," as they are brain dead but seemingly alive and hungry to spread their beliefs even though almost everything they say can be refuted. People (voters) who believe these things vote for the people who promise to put these lies into practice. The problem behind this phenomenon is not a lack of information. People who think this way do not have to do what they do.

It is their choice, and they may even be fully aware that their beliefs are irresponsible and irrational.

How can you cure this kind of false thinking and stubborn belief in things that are just factually wrong? How can you put an end to this creeping epidemic of insistence on obviously wrong facts by people who are not totally ignorant?

Thinking well means knowing and following the rules of rationality. It is not just a matter of having true beliefs. One must acquire the ability to distinguish bad arguments from good ones and then recognize how they support or refute a principle or point of view. A reasonable and responsible person will examine the evidence, and if it contradicts their beliefs and knowledge, they will act accordingly and reject the false or fabricated evidence as such. At a certain point, a kind of wisdom can be achieved because wisdom is the knowledge of one's own knowledge, i.e., self-knowledge. A wise person knows what he knows, and he makes sure that he is guided by his self-knowledge. In this way, the wise person is able to achieve what the Greeks called "eudaimonia," which is often, but incorrectly, translated as "happiness." It is, in fact, so much more.

Changing people is no easy feat, but there is no reason to think it is impossible. All they have to do is abide by the rules of thought. However, when unfounded, wrong, or even harmful ideas occupy the attention and energy of a growing part of humanity, it is up to all of us to show these people the habit of wrong thinking. At a time when humanity is threatened by great disasters, we must not display the arrogance of the wise to the willfully stupid. All people must be taken seriously, especially in dangerous times, because they are all afraid. They do not say it, but they feel it. Two German philosophers analyzed German fascism and commented on its alleged success as follows: "One of the lessons of the Hitler period is the stupidity of cleverness.... it is distinguished by well-informed superiority." (Horkheimer and Adorno, 2007)

The feeling of superiority has never done anyone any good, and we would do well to learn our lessons from history. Dividing society into good and bad, knowing and ignorant, or stupid and smart is not conducive to achieving our goals. We must take everyone with us on the path to a common *we*. That must be the goal. In doing so, those who are informed, clever, and perhaps even wise must support those who are so caught up in their fear of the future that they prefer to follow the pied pipers rather than the wise majority. We, as a whole, must become better at achieving our common goals. This has to do with the process of becoming more and more responsible for ourselves — and for everyone else at the same time. As Ludwig Hohl, an Austrian writer (1904–1980), said, "Thinking is above all courage."

Reference

Horkheimer, M. and Adorno T.W. (2007). *Dialectic of Enlightenment: Philosophical Fragments*.

Eudemonia

The Greek word "eudaimonia" is commonly translated as "happiness" or better: "well-being." For Aristotle (384–322 BC), the towering figure of ancient Greek philosophy, eudaimonia was the highest human good. Previously, the Greek poet Hesiod (c. 750–650 BC), relatively unknown to us, had called the word "sophrosyne" the golden rule for life. It means moderation or self-control, and its antonym, as we know, is "hubris." In his major work, the *Theogony*, Hesiod wrote about the Muses, all daughters of Zeus and Mnemosyne ("memory"). The Muses were personifications of knowledge and the arts, especially poetry, literature, dance, and music.

Two hundred years after Hesiod, Greek society had evolved. In Greek mythology, the Graces were the inspiring goddesses of grace, beauty, adornment, joy, gladness, festivity, dance, and song. On a Greek vase painting, Eudaimonia, the goddess of happiness and one of the younger muses, is depicted as one of the handmaidens of Aphrodite, the goddess of lust and pleasure.

Aristotle is the most famous philosopher of the classical period of Greek history. In his ethics, which Aristotle viewed as a practical rather than a theoretical study, he considered eudemonia — well-being — to be the highest goal of all conscious human action. To achieve this kind of happiness, one must have good character ("arete"), a kind of morality or ethical virtue. To achieve such a happy or virtuous character, one must have the right teachers and then be able to consciously choose to do the best possible in life. Only through this kind of behavior can one achieve the highest possible human virtue — becoming a wise person.

The question of the right way of life and "eudaimonia" is also discussed in Plato's dialogue "Gorgias." It says (Gorgias, sections 470e and 507c) that good souls are characterized by prudence and self-discipline, while bad souls are selfish and self-absorbed. The one who controls himself is virtuous. The prudent person is brave and just and therefore acts rightly, and only he lives a successful life in the world, in the state of eudaimonia. However, he who commits injustice acts unwisely; he does not subject the desires of his soul to the law of reason. This inevitably leads to his unhappiness. This is true even if he is outwardly successful and no one holds him accountable. Thus, for Plato, wrongdoers who escape punishment are even more unhappy than those who are punished.

In the dialogue, the interlocutor Callicles takes a radically anti-philosophical counter-position. Callicles does not consider the goal of lasting eudaimonia through prudence and restraint of desires to be worth striving for. For him, the suppression of desires is tantamount to a renunciation of life. For Callicles, eudaimonia is pleasure and the unrestrained enjoyment of desires. However, Callicles does not expect a permanent state of happiness as a result. From his point of view, a lasting eudaimonia is not desirable at all, because everything static is as dead as a stone for him. Callicles considers the philosophical striving for permanent eudaimonia to be fundamentally wrong since life can only be enjoyable through a constant alternation of pleasure and discontent.

The good life, well-being, and happiness have always been central themes for people. According to the ancient philosophers, well-being is based on ethics, in which happiness plays an important role. Happiness is considered an important part of human existence, but it can only be achieved on the basis of knowledge and understanding. Only when humans attain a certain level of wisdom can they be truly happy.

References

Krugman, P. (2008). The increasing returns revolution in trade and geography. Prize Lecture, December 8. https://www.nobelprize.org/uploads/2018/06/krugman_lecture.pdf

National Geographic. (n.d.). Greek philosophers. https://education.nationalgeographic.org/resource/greek-philosophers/

Platos. (380 BC). Gorgias. https://freeclassicebooks.com/Plato/Gorgias.pdf

Pursuit of Happiness. (n.d.). What Is Happiness? https://www.pursuit-of-happiness.org/what-is-happiness/

Stanford Encyclopedia of Philosophy. (n.d.). Aristotle's Ethics. https://plato.stanford.edu/entries/aristotle-ethics/

Stanford Encyclopedia of Philosophy. (n.d.). Happiness. https://plato.stanford.edu/entries/happiness/

Stevenson, B. and Wolfers, J. (2008). Economic growth and subjective well-being: Reassessing the Easterlin Paradox. Brookings Papers on Economic Activity, **2008**: 1–87. https://web.archive.org/web/20120617020438/http://bpp.wharton.upenn.edu/betseys/papers/happiness.pdf

Wikipedia. (n.d.). Eudaimonia. https://en.wikipedia.org/wiki/Eudaimonia

IV. Energy Solutions

Energy Technologies

E nergy — like happiness — has always been at the core of human society. It began with fire and its harnessing. Fire came from nature, and, at that time, what brought fire (e.g., lightning) was associated by people with God or heaven. The more energy human society could harness, the greater its prosperity became. Mathematically, one gallon of gasoline is equivalent to 450–500 hours of human labor. No wonder the hunger for energy remains at the heart of the human quest for prosperity.

One of the greatest challenges of the 21st century is the energy question. Today, the energy sector is the largest single source of carbon emissions in the world, accounting for about 41%. All fossil fuel-powered technologies are outdated, dirty, and ultimately a deadly trap for us and the planet. To reduce greenhouse gas (GHG) emissions, these technologies must be replaced with clean energy sources, such as wind, solar, and hydroelectric power, in order to provide a sustainable, livable future. This process of decarbonizing the Earth's atmosphere is only possible if the energy hunger of some ten billion people — most of them in the developing world — is satisfied and the need for clean energy for the increasingly energy-hungry industry is realized. How can this be achieved?

A rapid transition to renewable energies could be accomplished relatively quickly, both technically and financially, according to a study by an international group of researchers at Humboldt University in Berlin, published in the journal *Science*. According to the study, this transition would require only a fraction of the money that was spent on stimulus programs during the COVID-19 pandemic. Marina Andrijevic (Climate Analytics and Humboldt University of Berlin) said, "If just a fraction of this money was invested in climate-positive recovery plans, the world could achieve net zero carbon energy by mid-century. This is not about diverting money from COVID-19 stimulus or other low-carbon investments in industry, research, and development, but providing for the win-win solution of a boosted economy that simultaneously helps our efforts to stall climate change."

It will not be possible to radically stop the production and consumption of fossil fuels, but recent political developments show that continued dependence on oil, coal, and gas also has a political dimension that must be addressed. The task of transforming our global economy can be summed up in a simple picture: We need to completely transform our fast-moving, global, complex economic machinery on the fly, without stopping it or even causing it to stutter.

Innovative technologies are needed to achieve this, but these technologies already exist. We just need to apply these technologies in bigger and better ways. Besides, it is still "cheaper to save the planet than to ruin it" (Hal Harvey, climate policy expert).

References

Andrijevic, M., Schleussner, C.-F., Gidden, M., *et al.* (2020). COVID-19 recovery funds dwarf clean energy investment needs. *Science*, 370(6514): 298–300.

Energy Fundamentals. (n.d.). The history of the word "energy". https://home.uni-leipzig.de/energy/energy-fundamentals/01.htm

Green Transportation.info. (n.d.). How many days of human labor is in a gallon of gasoline? https://greentransportation.info/index.html

Hot or Cool Institute. https://hotorcool.org/

Inter-agency Task Force on Financing for Development. (2021). Financing for Sustainable Development Report 2021. https://freeclassicebooks.com/Plato/Gorgias.pdf

Pain, Stephanie. (2017). Power through the ages. *Nature*, November 29. https://www.nature.com/articles/d41586-017-07506-z

United Nations. (n.d.). What is renewable energy? https://www.un.org/en/climatechange/what-is-renewable-energy

Transitional Solutions

T he rapid transition from centralized and vulnerable to decentralized and resilient energy is one of the biggest challenges of the near future. As the economy continues to grow, the entire energy supply must be changed from fossil fuels to renewables. "Fundamentally, no terrestrial civilization can be anything else but a solar society dependent on the Sun's radiation" (Vaclav Smil). Globally, climate change is expected to increase the need for cooling and decrease the need for heating as temperatures rise. Renewable energy (solar, wind, hydro) is the first choice to satisfy this need, but there is concern that other, more stable energy supply systems will be additionally required.

The first and most important point is and remains the decarbonization of the power grid. This means replacing ALL fossil fuel power plants with solar, wind, hydro, and other zero-emission plants. The problem with most renewable energy sources is that they are relatively unreliable — it has to be windy, or the sun has to be shining. Other renewables, such as hydro-power or geothermal energy, will have to fill the potential gaps.

On the path to electricity as the most important global energy source under a sustainable energy scenario, many countries, including Germany, have opted for gas-fired power plants as one of the transitional solutions following the shutdown of nuclear and coal-fired power plants. Natural gas power plants generate electricity and waste heat. They emit only half as much CO_2 as coal-fired power plants, but other pollutants such as NOx, SOx, and other particulates are produced in smaller quantities. The largest gas-fired power plant in the world — natural gas is a byproduct of oil production — is the Jebel Ali Power and Desalination Plant in the United Arab Emirates (UAE), with an installed capacity of 8.6 gigawatts.

Worldwide, gas contributed about 6300 TWh (trillion watt-hours), or 24% of global electricity generation in 2020. Under the 2050 net-zero-emissions target, gas-fired power generation should increase in the short term and gradually displace dirty coal-fired generation, but it is then expected to gradually decline after 2030. In this future scenario, gas-fired power plants will produce about 90% less energy overall starting in 2040 than they did in 2020, and gas-fired plants will need to be retrofitted with carbon capture, utilization, and storage (CCUS) technologies to meet ambitious emissions targets. To keep them in operation while still meeting the levels within the net-zero scenario, natural gas as a fuel must then be successively replaced by carbon-free energy sources such as hydrogen.

References

Global Energy Monitor. (n.d.). Global oil and gas plant tracker. https://globalenergymonitor.org/projects/global-oil-gas-plant-tracker/

Global Energy Monitor. (2023). Gas plant capacity in development globally grew 22% in 2022. March 1. https://globalenergymonitor.org/press-release/gas-plant-capacity-in-development-globally-grew-22-in-2022/

International Energy Agency. (2022). Global Hydrogen Review 2022. https://www.iea.org/reports/global-hydrogen-review-2022

Office of Energy Efficiency & Renewable Energy. (2023). How much power is 1 gigawatt? August 24. https://climatechampions.unfccc.int/electrify-everything/

Wikipedia. (n.d.). List of power stations. https://en.wikipedia.org/wiki/Lists_of_power_stations

World Resources Institute. (n.d.). Global power plant database. https://datasets.wri.org/dataset/global-powerplantdatabase

Plan B: Nuclear Energy

Reliable energy in the form of electricity and heat could also come from a new generation of nuclear power plants — the small modular reactors (SMRs). The first of its kind has been in operation since May 2020 on the Akademik Lomonosov barge. Critics call Russia's first floating nuclear power plant "Chernobyl on ice."

SMR technology is diverse. Some reactors use the technology of today's light-water nuclear power plants. Another concept is high-temperature gas-cooled reactors (HTGRs), which operate at very high temperatures and use graphite as a coolant. Also, fast neutron reactors (FNRs) maintain the chain reaction by using fast neutrons, and there are also fast sodium reactors and molten salt reactors. These are all fourth-generation (Gen IV) nuclear reactors, while most of the 400 or so reactors currently in commercial use are second-generation reactors. SMRs are designed for series construction; several of them can be combined into a large power plant complex. Because of their relatively small size, SMRs can also be built in place of decommissioned coal- or gas-fired power plants.

In late 2021, the United States (US) and Romania signed the Cooperation Agreement 21–728 about US–Romania cooperation on SMRs. A 6-module NuScale plant is scheduled for completion in Romania by 2028. The project was announced on the sidelines of COP26 by John Kerry, the US Special Presidential Envoy for Climate, and Romanian President Klaus Iohannis. The new nuclear power plant will replace baseload power generation from coal-fired plants. NuScale, the US company building these SMRs, promotes its product as "safe, simple, small, economical, scalable nuclear power generation." Its promotional brochures tout the creation of numerous jobs, a boost to the economy, and increased tax revenues for the communities

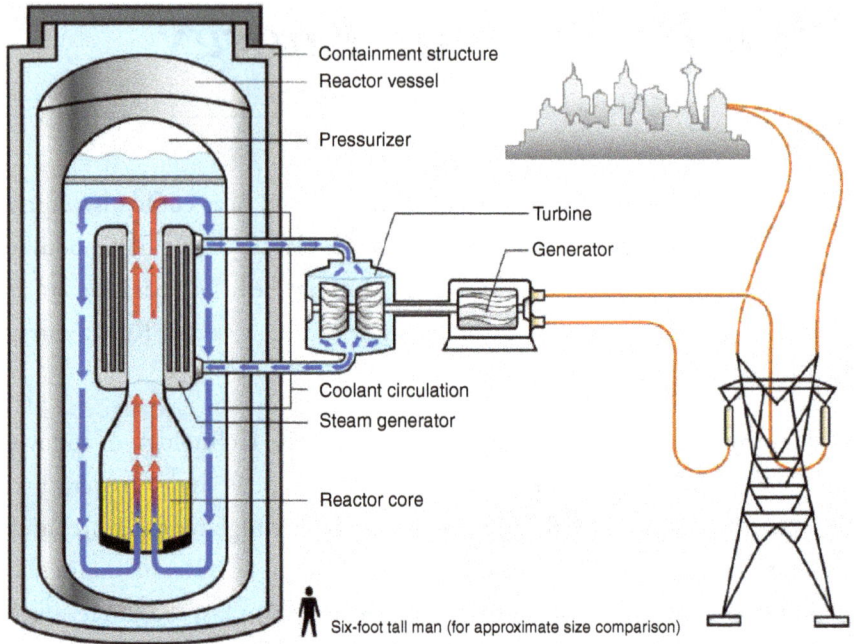

Containment structure
Reactor vessel
Pressurizer
Turbine
Generator
Coolant circulation
Steam generator
Reactor core
Six-foot tall man (for approximate size comparison)

Figure 25.   Illustration of a small modular reactor. (Source: US DoE)

where the nuclear plants are built. NuScale will build the first plant on US soil in Idaho Falls, and it already has agreements with other countries bordering Russia, such as Ukraine and Kazakhstan.

Although touted as safe, emission-free, and reliable, all nuclear power plants carry the risk of accidents or "incidents." Three Mile Island (Harrisburg), Chernobyl (Ukraine), and Fukushima (Japan) are only the worst disasters in a long list of nuclear power plant failures worldwide. Nuclear power plants also pose high risks in the event of war (e.g., Ukraine).

The waste from nuclear power plants is certainly the biggest problem to address. There are no long-term solutions for nuclear waste, which remains radioactive for at least a hundred thousand years. There are several research projects in the pipeline to make use of nuclear waste material, but none have yet been successfully tested. The entire life cycle of nuclear power plants, the mining of uranium ore, the cancer risk for the people handling the material, severe safety risks during the life of the plant, and the possible use of the waste as "nuclear weapons" all make nuclear power plants a tempting but very dangerous solution to mankind's energy problem.

Nuclear power plant operators make a simple calculation: The energy they can sell is an asset, something counted as property on their books. The companies do not want to write off their assets, and from a capitalist point of view, why should they? Just to save the world? That is why many of today's politicians want to keep old and unsafe nuclear power plants running. They like to talk about the high reactor safety and the cheap, reliable electricity that nuclear power plants supposedly provide. Potential risks are downplayed while the real costs of nuclear power plants, especially for safely storing nuclear waste for tens of thousands of years, are kept quiet.

The European Union (EU) has created a set of rules for sustainable investments — "an EU-wide classification system for sustainable activities," also called the "EU taxonomy." In early 2022, the EU Commission added nuclear- and gas-powered plants to the list of sustainable energy sources. Despite the many deadly, costly, and devastating disasters caused by nuclear accidents, similar power plants are now called "sustainable" energy sources. It seems that lobbying and national interests also apply to the term "sustainability" — not only is John Kerry a supporter of SMR technology, but he is also the US Special Presidential Envoy for Climate. The battle for sustainability has begun. It will be a long and difficult fight, and no one knows what this particular conflict will bring.

Nuclear Fusion

---

nother option for future energy production is nuclear fusion. The process in such reactors appears to be quite simple: Two light atomic nuclei [deuterium (H-2) and tritium (H-3)] fuse to form a heavier nucleus (He-4), releasing a lot of energy. This process is similar to energy production in stars such as our Sun. The hard part is creating a deuterium-tritium plasma in some kind of giant magnet at a working temperature of more than one hundred million degrees Celsius. The real problem at the moment is that you have to put more energy into such a reactor system than you get out.

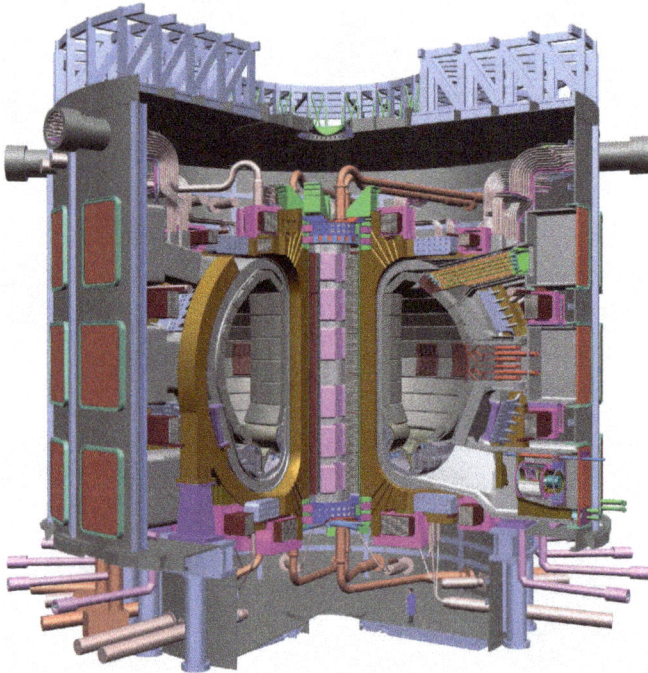

Figure 26.   Concept of a fusion reactor. (Source: US DoE (9786811206))

The International Thermonuclear Experimental Reactor (ITER) project is one such nuclear fusion reactor that is currently under construction. It is a tokamak reactor, the most advanced of several possible fusion reactor designs. The heart of a tokamak is its donut-shaped vacuum chamber. ITER is an international project that was launched in 1985. The 27 countries of the EU are involved, as well as (through Euratom) Switzerland and the United Kingdom (UK), plus China, India, Japan, Korea, the Russian Federation, and the US. ITER is expected to generate 500 megawatts (MW) of energy from 50 MW of added thermal energy. The timetable calls for the machine to first come online for a test run in 2025, but this plan, like many others before, had to be postponed without a new deadline established. The original budget for ITER was just under 6 billion euros; the total investment is likely to exceed 22 billion euros, although some fear it will be closer to 40 billion euros. ITER is one of the most ambitious and complicated engineering projects in human history. In any case, the cost is very, very high, and if the plant really works, which may happen around 2035, the project duration from concept to completion is far too long at 50 years. The current need to rapidly move away from fossil fuels will not be met with nuclear fusion reactors.

Bill Gates, founder of Microsoft and renowned philanthropist, is partnering with Commonwealth Fusion Systems (CFS) to support the development of a novel magnet technology at MIT. The first successful operation of a prototype took place in December 2021, and MIT scientists already see it as a breakthrough for a new, clean, sustainable, and always available energy source. Other investors in the project include Jeff Bezos, Richard Branson, and Ray Dalio with their Breakthrough Energy Ventures fund. The fund aims to accelerate the energy transition across all sectors of the economy. Other projects supported by the fund include ZeroAvia (the first zero-emission propulsion system for aviation), C-Zero (decarbonization of natural gas), and Electric Hydrogen (production of green hydrogen from water and renewable energy).

The diversified funding system of these and other entrepreneurs could lead to the goal faster than other multinational collaborations, such as ITER with its project duration of 50 years. However, there is a long way between the successful test run of a single magnet and a functioning, reliable, and safe machine.

References

Breakthrough Energy. https://breakthroughenergy.org/

Chatzis, I. and Barbarino, M. (n.d.) What is fusion, and why is it so difficult to achieve? International Atomic Nuclear Agency. https://www.iaea.org/bulletin/what-is-fusion-and-why-is-it-so-difficult-to-achieve

European Commission. (n.d.). EU taxonomy for sustainable activities. https://finance.ec.europa.eu/sustainable-finance/tools-and-standards/eu-taxonomy-sustainable-activities_en

ITER. https://www.iter.org/

Liou, J. (2023). What are small modular reactors (SMRs)? International Atomic Energy Agency, September 23. https://www.iaea.org/newscenter/news/what-are-small-modular-reactors-smrs

McDonald, S.M. (2021). Is nuclear power our best bet against climate change? *Bosten Review*, October 12. https://www.bostonreview.net/articles/is-nuclear-power-our-best-bet-against-climate-change/

MIT Energy Initiative. (n.d.). Membership. https://energy.mit.edu/member/cfs/

Nuclear Energy Agency. (n.d.). Nuclear Data Services. https://www.oecd-nea.org/jcms/pl_27360/nuclear-data-services

Nuclear Energy Institute. (2020). Nuclear by the numbers. August. https://www.nei.org/CorporateSite/media/filefolder/resources/fact-sheets/nei-nuclear-by-the-numbers-092520-final.pdf

NuScale Power. https://www.nuscalepower.com/en

Office of Nuclear Energy. (n.d.). Advanced small modular reactors (SMRs). https://www.energy.gov/ne/advanced-small-modular-reactors-smrs

The White House. (2021). Fact sheet: President Biden tackles methane emissions, spurs innovations, and supports sustainable agriculture to build a clean energy economy and create jobs. November 2. https://www.whitehouse.gov/briefing-room/statements-releases/2021/11/02/fact-sheet-president-biden-tackles-methane-emissions-spurs-innovations-and-supports-sustainable-agriculture-to-build-a-clean-energy-economy-and-create-jobs/

US Global Leadership Coalition. (n.d.). John Kerry, Special Presidential Envoy for Climate. https://www.usglc.org/positions/special-presidential-envoy-for-climate/

Wikipedia. (n.d.). Fusion power. https://en.wikipedia.org/wiki/Fusion_power

Renewable Energies

R enewable energies (with the exception of tidal power plants) are ultimately all provided by our sun. The Sun radiates 174,423,000,000 kilowatt-hours of energy onto the Earth per hour. About 60% of this amount reaches the Earth's surface. The total energy consumption on the Earth in 2017 was 113,000,000,000 kilowatt-hours. This means that solar radiation provides more energy per hour than the entire world consumes per year. Yes, that's right — the amount of solar energy hitting the Earth every hour is enough to power our planet's population for an entire year!

In order to achieve the goals of the 2015 Paris Agreement, renewable energies will have to cover the entire energy demand of the planet very soon, with some exceptions, such as fuel for airplanes. This is estimated to require investments of around $1.4 trillion per year. This amount is needed to restructure the energy sector, to give an idea of the true cost of a Net Zero world by 2050. However, according to research, annual government subsidies to the fossil fuel industry amount to roughly $5 trillion per year; this money could and should go into renewables!

Renewable energy is defined as energy that can be renewed, as opposed to fossil fuels such as gas, oil, or coal that can only be burned, releasing a lot of CO_2 in the process. Renewable energy sources, such as solar, wind, hydro, and biofuels, together account for about 23% of the world's electricity generation today. Clearly, to meet the net-zero scenario's timetable of 60% by 2030, renewables will have to continue to expand very, very quickly.

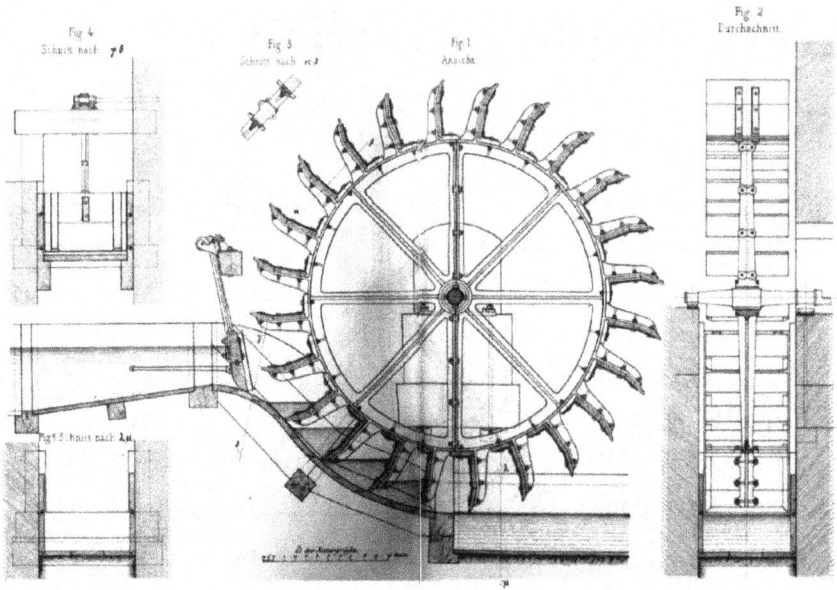

Figure 27.   Theory and construction of water wheels, Redtenbacher, Ferdinand (1809–1863). (Public Domain)

The use of renewable energy began over 2,000 years ago with water wheels. A water wheel uses the energy of moving water to drive a machine connected to it by the wheel's rotating shaft. Beginning in the seventh century AD, windmills were used to power other renewable energy devices. By 1600, windmills were ubiquitous in northern Europe, especially in the Netherlands. Windmills were used primarily to pump water and grind grain. Solar energy was used by Archimedes, a Greek mathematician and inventor. In 212 BC, Roman ships that invaded Syracuse were set on fire with mirrors that captured and channeled the power of the sun. Leonardo da Vinci also made drawings of concave mirrors, though there is no evidence that machines were built based on these drawings.

Unlike other energy sources, the generation and use of renewable energy have been increasing for some time, and investor interest in renewable energy remains strong. Renewable energy equipment manufacturers and project developers are operating in a healthy business environment thanks to expectations of a steady growth in demand. In 2021 alone, there was a huge upswing in renewable-energy-capacity construction of about 10%. In the coming years, cost reductions and continued political support — also in light of the Ukraine war, which made dependence on unreliable partners such as Russia impossible — should enable strong growth in renewables. If forecasts are to be believed, renewables will replace coal as the largest source of electricity generation in 2025. An important aspect for further accelerating the use of renewables as the "energy of the future" is the net-zero emission targets in key markets like the EU, Japan, South Korea, and China.

References

Encyclopaedia Britannica. (n.d.). Waterwheel. https://www.britannica.com/technology/waterwheel-engineering

European Environment Agency. (2022). A future based on renewable energy. November 28. https://www.eea.europa.eu/signals-archived/signals-2022/articles/a-future-based-on-renewable-energy

International Energy Agency. (2023). World Energy Investment 2023. May. https://www.iea.org/reports/world-energy-investment-2023

International Renewable Energy Agency. (n.d.). Investment. https://okmagazine.com/p/matthew-perry-relaxed-friend-final-public-sighting-tragic-death/

International Renewable Energy Agency. (2023). World Energy Transitions Outlook 2023: 1.5°C Pathway. June. https://www.irena.org/Publications/2023/Jun/World-Energy-Transitions-Outlook-2023

National Grid. (n.d.). How will our electricity supply change in the future? https://www.nationalgrid.com/stories/energy-explained/how-will-our-electricity-supply-change-future

Wikipedia. (n.d.). Windwill. https://en.wikipedia.org/wiki/Windmill

Wisconsin Center for Environmental Education. (2020). Facts about future energy resources. https://www3.uwsp.edu/cnr-ap/KEEP/Documents/Activities/Energy%20Fact%20Sheets/FactsAboutFutureEnergyResources.pdf

History of Solar Energy

The Sun is the most important source of energy on Earth. It is clean, abundant, and, in regions with plenty of sunshine, reliable. Solar energy systems convert sunlight into electrical (and thermal) energy. This can be done using a variety of methods. There are photovoltaic (PV) cells, solar thermal collectors, and Concentrated Solar Power (CSP) systems. These use solar energy to heat water or generate steam, which is then converted into electricity by a turbine. Since solar collectors operate only during the day, large energy storage capacities are required.

In ancient Greece, a method of lighting fires with magnifying glasses or mirrors was used from about the seventh century BC. These mirrors were called "burning mirrors (or glasses)." The lighting of the Olympic torch was done at Olympia in Greece with such a burning glass. In his *Naturalis Historia* (Natural History), the Roman writer Pliny the Elder (c. 23–79 AD) described convex lenses used to cauterize (in diathermy, "close") wounds. At about the same time, burning glass devices were already being used in China. In architecture, the ubiquitous Roman bathhouses had large, south-facing windows to take advantage of the Sun's heat. The Anasazi, the ancestors of the Pueblo population in what is now the US, also lived in south-facing cliff dwellings that used the winter sun as a natural heater.

In 1767, Horace Bénédict de Saussure (1740–1799), a Swiss scientist and one of the most important European natural scientists of his time, invented the precursor of our modern solar collector. In 1839, French scientist Edmond Becquerel (1820–1891) observed the release of charge carriers (ions) when light strikes a metal surface (the "Becquerel effect"). Heinrich Hertz (1857–1894), a German physicist, was the first to observe the photoelectric effect in 1886. This is practically the basis of the PV effect, which enables the conversion of light energy into electrical energy. Today's solar cells use this effect to convert sunlight into electricity. However, it took a long time to progress to the modern solar collector with relatively high efficiency.

In 1866, Augustin Mouchot, a French teacher, developed the first parabolic trough solar collector. He connected a container filled with water to the collector and used the accumulated steam to power a small engine. Mouchot's "solar engine" ran without fuel; it was powered only by the Sun's rays, and this had a bizarre, magical, and even chilling effect on observers. William Adams, a British engineer who worked at the British Patent Office in London, wrote of Mouchot's discoveries: "This idea may be, and probably is, purely Utopian, but very important discoveries have been made in striving for the impossible; and if no further success is achieved than that of utilizing the rays of the sun for driving stationary steam engines, an important addition to physical science will have been made, and a great commercial revolution will have been effected." However, when imported coal from the UK became cheaper, and coal was additionally found in eastern France, it was clear that the days of this "magic" machine were numbered.

Figure 28. Parabolic solar collector, Augustin Mouchot, 1878. (Public Domain)

In the US, during the Second Industrial Revolution, many inventors and businessmen tried to harness and market the power of the sun. The first solar cell, or PV cell, was constructed by Charles Fritts (1850–1903) in 1883. It produced an electric current "continuously, constantly, and with considerable power." Fritts' solar cell had an efficiency of only 1–2%, but it still marked the beginning of modern PV technology. The vast majority rejected Fritts' invention because it had no moving parts and used no fuel. People did not understand it at all, and some even dismissed it as a completely unscientific "perpetual motion machine." In 1888, inventor Edward Weston (1850–1936) received two patents for solar cells "for converting the radiant energy given off by the sun into electrical energy." In 1909, Wiliam J. Bailey of the Carnegie Steel Company patented and built solar collectors — called "solar heaters" — with a flat plate collector and a separate storage tank. These systems were sold in large numbers in California and Florida. Cheap fossil fuels ended this boom of the early solar panel industry in the 1930s.

One thing I feel sure of, and that is that the human race must
finally utilize direct sun power or revert to barbarism.

— Frank Shuman, 1914

In 1910, American entrepreneur Frank Shuman (1862–1918) built a solar array in his backyard in Tacony, Philadelphia. It had a total area of about 10,000 square feet and generated approximately 25 horsepower. To make his technology a real contender in the energy and power market, he needed to build much larger systems. Shuman was certain about the potential success of his machine: "The future development of solar power has no limit. Where great natural waterpower exists, sun power cannot compete; but sun-power generators will, in the near future, displace all other forms of mechanical power over at least 10 percent of the earth's land surface; and

FORM OF MACHINE THAT PUTS THE RAYS OF THE SUN TO WORK DIRECT.

Figure 29.   Shuman's solar engine, 1907. (Public Domain)

in the far distant future, natural fuels having been exhausted, it will remain as the only means of existence of the human race." (Shuman, 1911, S. 291).

To move forward with his plans, Shuman needed to find investors. After he failed to do so in the US, he went to London. Here, he found partners with whom he founded the Sun Power Company. Shuman wanted to go to Egypt (then a British protectorate) because there was plenty of sunshine and an existing mechanized irrigation system on which he could build his solar system. Shuman leased land in Maadi, then a suburb of Cairo and the administrative center of Egypt. With British financial support and British physicist Sir Charles Vernon Boys (1855–1944) as a scientific advisor, the design of the system was updated and completed in June 1913. Shuman's system bears only a vague resemblance to today's solar power plants. Even so, Shuman was certain that he had a technology in hand that would convince not only his backers but the entire world that solar power was an excellent alternative to generating electricity from coal.

As always, history took its own course. With the outbreak of World War I, Shuman's plans for a larger plant — financed by German backers — were scuttled. The Maadi plant was dismantled to provide parts and scrap metal for the war effort. In 1918, Shuman died of a heart attack. After the end of World War I, fossil fuels, especially gasoline, became cheaper. As a result, solar technology became economically uncompetitive, so further development of solar technology came to a standstill for a long time.

It was not until the 1950s that PV technology was further improved upon at Bell Laboratories (US). In 1954, three physicists (Daryl Chapin, Calvin Fuller, and Gerald Pearson) invented the first practical solar cell with an efficiency of up to 6%. The new technology was still too expensive for home use, but not for space. Vanguard 1 was the first satellite to use solar cells to partially power a spacecraft. In 1964, the Nimbus satellite was launched by NASA and powered entirely by its 470-watt PV solar array. Since then, solar power has been used extensively in space and gradually deployed on Earth.

The oil price shock of the 1970s brought solar energy back to the center stage. In 1974, the US Congress passed the Solar Energy Research, Development and Demonstration Act with the goal of "making solar energy viable and affordable for public use." After some difficulties in the years before 2000, solar energy is now booming, with an average growth of up to 40–50% annually. In the US alone, more than 100 gigawatts (GW) of solar capacity is currently installed. New-age solar companies such as Tesla now offer better-designed and more advanced systems called "building-based photovoltaics," in which PV cells are integrated into roof tiles or glass facades.

Solar thermal power plants use a large number of mirrors to concentrate sunlight onto a radiation collector or absorber, usually mounted high on a tower. The energy then heats some kind of medium, usually thermal oil, water/steam, molten salt, or air. The resulting thermal energy is converted into electricity by a turbine. The largest of the many challenges of this high-temperature technology is accurately directing solar radiation to the absorbers, avoiding heat loss, and keeping costs down. As always, this is not easy when taking the first steps in a new and challenging technology. There are three different concepts: Linear Fresnel Collector (LFC) plants, Parabolic Trough Collector (PTC) plants, and Concentrated Solar Power (CSP) plants. The primary product of these solar thermal plants is heat that can be stored, converted, or used on-site as industrial process heat. In the solar thermal MATS (Multipurpose Applications by Thermodynamic Solar) plant in Egypt, thermal energy is also used to desalinate seawater. The latter two technologies (PTC and CSP) dominate the growing global market for solar thermal power plants to meet rising energy demand, particularly in countries of the Global South.

After decades of research and development aimed at making electricity from solar thermal power plants cheaper, a growing number of such plants are now operating commercially. Nevada Solar One was a large pilot project in the Mojave Desert in the US. It was built by the DoE (Department of Energy) in cooperation with a number of US utilities and began operation in 1981. The plant generated only 7 MW of electricity. In 1995, the plant was expanded. Solar Two had a capacity of 10 MW. Its success was an important milestone for the development of much larger solar energy projects.

The Ouarzazate solar power plant in Morocco, also known as the Noor power plant, is one of the largest solar power plants in the world (as of 2022). The Ouarzazate complex is a solar park that includes several solar power plants using different solar technologies. The total capacity of the power plant will be 580 MW with a storage capacity of 7 to 8 hours. The capacity of a typical nuclear power plant is about 1,000 MW; Noor has more than half that capacity.

The "Noor" is the flagship project of Morocco's ambitious energy policy, which aims to increase the share of renewable energy to over 50% by 2030. The Noor solar power plant in Ouarzazate produces carbon-free energy equivalent to 2.5 million tons of imported oil per year. This plant alone (Noor1-4) reduces carbon dioxide emissions by nearly 800,000 tons per year, which is nearly 5% of Morocco's total GHG emissions. It also offers the possibility of exporting green energy to neighboring countries, contributing to the socio-economic and cultural development of the entire region.

The United Arab Emirates (UAE), a federation of seven emirates, is taking major steps to develop solar energy as its main energy source. In Abu Dhabi, the Shams Solar Power Plant, a 100-megawatt power plant using CSP technology, was commissioned in 2013. The Noor Abu Dhabi solar power plant in Sweihan, Abu Dhabi, went online in April 2019. It is the world's largest solar power plant, with a capacity of 1.2 GW. In 2020, a 2 GW PV project, Al Dhafra Solar, was announced. This plant will be the largest PV power plant in the world when completed. Dubai (another member of the UAE) is targeting 75% renewables, mainly solar, in its energy mix by 2050.

The UAE has already begun construction of the first solar-powered green hydrogen plant in the Middle East and North Africa — the Mohammed bin Rashid Al Maktoum Solar Park, a 2.3 billion euro "test bed for green hydrogen production." The project is part of the UAE's Net Zero by 2050 strategic initiative. The UAE's goal is to achieve a 25% share of the global hydrogen market by 2030. With the energy market changing rapidly, partly due to the war in Ukraine, major consumers such as Germany are already knocking on the door of projects such as this one to help secure green energy supplies for the future.

The Saudi Arabian oil state is also building a huge technology and energy park. NEOM ("New Future") is an eerie showcase project designed from scratch. It is being built on the Red Sea, close to the border with Egypt, on the initiative of Saudi Crown Prince Mohammed bin Salman and with a planned investment volume of US$500 billion. The project is intended to diversify Saudi Arabia's oil economy for a post-oil era. In 2022, the foundation stone was laid for a plant to produce green hydrogen. With the involvement of a subsidiary of ThyssenKrupp, the hydrogen is to be exported via the port of Duba starting in 2026.

The largest operating solar thermal power plant in the US is the Ivanpah Solar Power Facility in the Mojave Desert (California). The first unit of the plant was commissioned in 2014, and at that time, it was the largest operating solar thermal power plant in the world, with a capacity of 392 MW. It cost US$2.2 billion.

The cost of building and installing PV systems is falling; the technologies for building ever-larger solar power plants are now absolutely reliable. In addition, solar power prices are now so competitive that solar power

Figure 30.   Ivanpah Solar Electric Generating System. (Source: Sandra Ware — U.S. Department of Energy, https://commons.wikimedia.org/wiki/File:Ware_000605_170326_515119_4578_(36965579485).jpg)

generation has the potential to grow rapidly and on a large scale, providing an important piece of the sustainable energy mix. The construction of huge multi-terawatt solar power plants will be an essential part of the global strategy to mitigate the climate crisis.

How the use of solar energy can work on a much smaller and more practical level is shown by Shamsina. This Egyptian start-up uses simple technology to produce solar water heaters for low-income communities. Harvard University's Innovation Lab (BuildIt program) is supporting the project, founded by two Egyptian-American entrepreneurs. The invention is intended to help Egyptian families who currently lack the means to reliably and safely heat domestic water. The idea came about when the founders realized that families were using kerosene lamps, gas tanks, and open fires on a daily basis to heat water for their basic needs, such as bathing, cooking, and cleaning. These energy sources are very dangerous due to their negative effects on the health of the population (respiratory diseases), and as fossil fuels, they produce a lot of carbon dioxide that is harmful to the climate. The Shamsina project is proof that there is an ever-increasing number of ideas, inventions, and products around the world that will help to achieve the goal of realizing a carbon-free world.

References

Advancing Physics. (2009). This month in physics history. APS News, 18(4). https://www.aps.org/publications/apsnews/200904/physicshistory.cfm

Congress.Gov. (2019). H.R.3597 — Solar Energy Research and Development Act of 2019. https://www.congress.gov/bill/116th-congress/house-bill/3597/text

Emirates Water and Supply Company. (n.d.). Al Dhafra Solar PV. https://www.ewec.ae/en/power-plant/al-dhafra-solar-pv

Emirates Water and Supply Company. (n.d.). Noor Abu Dhabi. https://www.ewec.ae/en/power-plants/noor-abu-dhabi

European Commission. (2011). Multipurpose applications by thermodynamic solar. July 19. https://cordis.europa.eu/project/id/268219

GovTrack. (1974). H.R. 16319 (93rd): Solar Energy Research, Development and Demonstration Act. August 7. https://www.govtrack.us/congress/bills/93/hr16319

Green Technologies: Solar Water Heating. https://greentechswh.weebly.com/

Jones, G.J. and Bouamane, L. (2012). "Power from sunshine": A business history of solar energy. Harvard Business School, Working Paper 12-105, May 25. https://www.hbs.edu/ris/Publication%20Files/12-105.pdf

NEOM. https://www.neom.com/en-us

Office of Energy Efficiency & Renewable Energy. (2023). Making clean energy more accessible and affordable. January 26. https://www.energy.gov/eere/articles/making-clean-energy-more-accessible-and-affordable

Perlin, J. (2004). The silicon electric cell turns 50. National Renewable Energy Laboratory. https://www.nrel.gov/docs/fy04osti/33947.pdf

Platzer, W.J., Heimsath, A., Cuevas, F., et al. (2014). Linear fresnel collector receiver: Heat loss and temperatures. Energy Procedia. https://www.academia.edu/17781921/Linear_Fresnel_Collector_Receiver_Heat_Loss_and_Temperatures

Shamsina. https://shamsinasolar.com/

Shere, J. (n.d.). Frank Shuman's Solar Arabian Dream. Renewable. https://renewablebook.wordpress.com/chapter-excerpts/350-2/

Shuman, F. (1911). Power from sunshine: A pioneer solar power plant. Scientific American, September 30. https://www.scientificamerican.com/article/power-from-sunshine/

The Electrical Experimenter. (1916). The utilization of the Sun's energy. March. https://upload.wikimedia.org/wikipedia/commons/thumb/c/c2/The_Electrical_Experimenter%2C_Volume_3.pdf/page643-1181px-The_Electrical_Experimenter%2C_Volume_3.pdf.jpg

US Department of Energy. (n.d.). Ivanpah. https://www.energy.gov/lpo/ivanpah

US Department of Energy. (n.d.). The history of solar. https://www1.eere.energy.gov/solar/pdfs/solar_timeline.pdf

Wikipedia. (n.d.). Charles Fritts. https://en.wikipedia.org/wiki/Charles_Fritts

Wikipedia. (n.d.). Edward Weston (chemist). https://en.wikipedia.org/wiki/Edward_Weston_(chemist)

Wikipedia. (n.d.). Ivanpah Solar Power Facility. https://en.wikipedia.org/wiki/Ivanpah_Solar_Power_Facility

Wikipedia. (n.d.). Mohammed bin Rashid Al Maktoum Solar Park. https://en.wikipedia.org/wiki/Mohammed_bin_Rashid_Al_Maktoum_Solar_Park

Wikipedia. (n.d.). Nevada Solar One. https://en.wikipedia.org/wiki/Nevada_Solar_One

Wikipedia. (n.d.). Noor Abu Dhabi Solar Power plant. https://en.wikipedia.org/wiki/Noor_Abu_Dhabi_Solar_Power_plant

Wikipedia. (n.d.). Ouarzazate Solar Power Station. https://en.wikipedia.org/wiki/Ouarzazate_Solar_Power_Station

Wind Power

W indmills are one of the earliest technologies that converted the power of nature into a physical force by converting wind energy into rotational energy. This energy can be used for practical tasks such as grinding grain or irrigation. Heron of Alexandria (Greek mathematician and engineer, c. 10–70 AD) invented the windmill as early as the first century AD. His invention was intended to power an organ, but most of his further work has not survived. The oldest known windmill in practical use was built in Persia around 700 AD. This type of windmill (the "panemone," or horizontal-axis windmill) was widely used in Persia, India, and China.

Figure 31.   "Panemone," a Persian windmill. (Source: Hasanahmadifard, CC BY-SA 4.0 DEED)

Horizontal windmills were first used in northern Europe at the end of the 12th century and were mainly used for grinding grain. They became a typical symbol of the Netherlands. There, windmills were used to pump water from the lowlands into the rivers and ponds beyond the dikes so that the land could be farmed. Even today, there are more than 1,000 operational windmills in the Netherlands. Worldwide, they reached their peak of popularity around 1850, when approximately 200,000 were in operation. The Industrial Revolution put an end to most of them when steam engines, fueled by coal, became the main prime movers.

As early as 1887, the first wind turbines were built. In July 1887, James Blyth (1838–1906), a professor at Anderson's College, constructed a wind turbine to generate electricity for his cottage that ran for more than 20 years. That same year, Charles F. Brush (1849–1929), one of the founding fathers of the American electrical industry, built a huge windmill about 17 meters in diameter. The fully automated apparatus was used to feed the batteries the inventor had installed in his house to supply electricity. In 1890, *Scientific American*, an American popular science magazine, wrote about the machine: "It is found after continued use of this electric plant that the amount of attention required to keep it in working condition is practically nothing. It has been in constant operation more than two years, and has proved in every respect a complete success."

Poul la Cour (1846–1908), a Danish teacher and inventor, can rightly be called the "father of modern wind turbines." La Cour discovered that wind turbines with fewer rotor blades could turn faster and thus generate electricity more efficiently. La Cour used the electricity generated by his wind turbines to produce hydrogen for his school's gas lighting through the use of electrolysis. In 1903, he founded the "Society of Wind Electricians" as a platform for promoting rural electrification. When he died on April 24, 1908, there were 72 rural electric utilities in Denmark operating wind turbines ranging from 5 to 25 kW. Small windmills were

Figure 32. The Giant Brush Windmill in Cleveland, Ohio, 1998. (Public Domain)

also widely used in rural areas of the United States (US), where there was no electrical grid in the early 20th century, to generate electricity, for example, for a ranch.

In 1931, a vertical-axis wind turbine was patented by Georges Jean Marie Darrieus, a French aeronautical engineer. This type is still used today, but not as widely as horizontal wind turbines. The largest horizontal turbine was built in the same year, 1931, in Yalta, Crimea, and had an output of 100 kW. It featured a 32-meter tower and an energy efficiency of 32%, which is similar to today's wind turbines. In 1941, the first megawatt-class wind turbine was connected to the local power distribution grid at Grandpa's Knob in Castleton, Vermont, US. Although it failed after 1,100 hours of operation, it remained the largest wind turbine ever built until 1979.

Since the times of Poul la Cour, Denmark has been a pioneer in wind energy. The country sought early on to become independent of energy imports, such as foreign oil or natural gas, by promoting alternative energies. In 1956, Johannes Juul, a former student of Poul la Cour, built a 200-kilowatt wind turbine in Gedser, Denmark. The turbine was in operation until 1967, and Juul's invention, the aerodynamic emergency brake, is still in use today.

However, the true renaissance of wind turbines began after the oil shock of the 1970s. At that time, it became clear that there could be a shortage of fossil fuels and that oil and gas prices would continue to rise. Consequently, in the 1980s, large-scale industrial production of wind turbines began, with the first designs coming from Denmark (Vestas) and Germany (Siemens). Wind farms consisting of up to 1,000 individual wind turbines were then installed in Europe, the US (California), and elsewhere in the world. More recently, offshore wind farms have been favored by both developers and the public.

Wind turbines have subsequently become larger and are placed in locations where there is a steady wind. Modern horizontal-axis wind turbines (HAWTs) make up the majority of turbines. They are highly efficient, reliable, and run on a free, clean, renewable fuel — wind. According to statistics published by the World Wind Energy Association (WWEA), the total capacity of all wind farms worldwide has now reached 744 gigawatts, or about 7% of the world's electricity demand.

References

Danish Wind Industry Association. (n.d.). A wind energy pioneer: Charles F. Brush. http://xn--drmstrre-64ad.dk/wp-content/wind/miller/windpower%20web/en/pictures/brush.htm

Ohio Memory Collection. (n.d.). Scientific American. https://ohiomemory.org/digital/collection/p267401coll36/id/19823

Poul la cour Museet. https://www.poullacour.dk/en/home/ (in Danish)

Shubov, M. (2021). History of prime movers and future implications. ArXiv, April 19. https://arxiv.org/abs/2104.08981

Wikipedia. (n.d.). Charles F. Brush. https://en.wikipedia.org/wiki/Charles_F._Brush

Wikipedia. (n.d.). Darrieus wind turbine. https://en.wikipedia.org/wiki/Darrieus_wind_turbine

Wikipedia. (n.d.). Georges Darrieus. https://de.wikipedia.org/wiki/Georges_Darrieus

Wikipedia. (n.d.). Hero of Alexandria. https://en.wikipedia.org/wiki/Hero_of_Alexandria

Wikipedia. (n.d.). Panemone windmill. https://en.wikipedia.org/wiki/Panemone_windmill

Wikipedia. (n.d.). Wind power. https://en.wikipedia.org/wiki/Wind_power

World Wind Energy Association. https://wwindea.org/

Hydropower

ydropower is one of the oldest and most successful renewable energy systems in human history. There are horizontal and vertical water wheels. A horizontal wheel driven by a jet of water can directly set a millstone in motion. Vertical water wheels are more powerful and have two basic designs: undershot and overshot. The undershot wheel is a paddle wheel that rotates under the momentum of a water stream; when water becomes scarce, the power decreases accordingly. Overshot water mills are driven by water coming from above, either through specially designed channels or through an elevated water inlet.

The first mention of water wheels dates back to ancient Greece, where they were used for grinding wheat. The Romans then adopted the Greek design. Water mills were used throughout the Roman Empire and are described in the works of Vitruvius (25 BC) and Pliny the Elder in his *Naturalis Historiæ* (77 AD). An exceptional example is the mill of Barbegal, located 12 km north of Arles in France. Dating back to the 4th century AD, the mill had 16 water wheels, and its production capacity was estimated at about 4.5 tons of flour per day.

Water wheels were mainly used for grinding grain but could also be employed for sawing marble. When Rome was besieged in 537 AD, the warlike Goths cut off the city's water supply through the aqueducts. Belisarius, the general defending the city, then had floating mills built. The construction used two rows of boats with water wheels suspended between them. The new design worked so well that it was soon copied throughout Europe. In China, water mills have been used since at least 31 AD, when engineer Tu Shih invented a water-powered machine for making agricultural iron equipment.

Since the 9th century, large parts of Europe have been farmed by monastic communities. Cereals were an important crop, and most of the harvest was ground with water mills. In the 9th century, the abbey of Saint-Germain-des-Prés in France, one of the great abbeys between the Loire and the Rhine, had a total of 84 water wheels. In England, after the Conquest, the Norman King William had all his possessions counted, including the country's mills. The *Domesday Book* of 1086 lists more than 5,600 mills. These mills were powered either by water or livestock and were used for a variety of purposes. In 1098, the Cistercian Order was founded. It was one of the many medieval orders (Benedictines, Carthusians), but typical for the Cistercians was communal living. Cistercian monks lived in well-organized monasteries located near rivers that provided water, power, and sewage. A typical Cistercian monastery was located on a millrace (an artificial watercourse) that powered a water mill for grinding wheat and other grain.

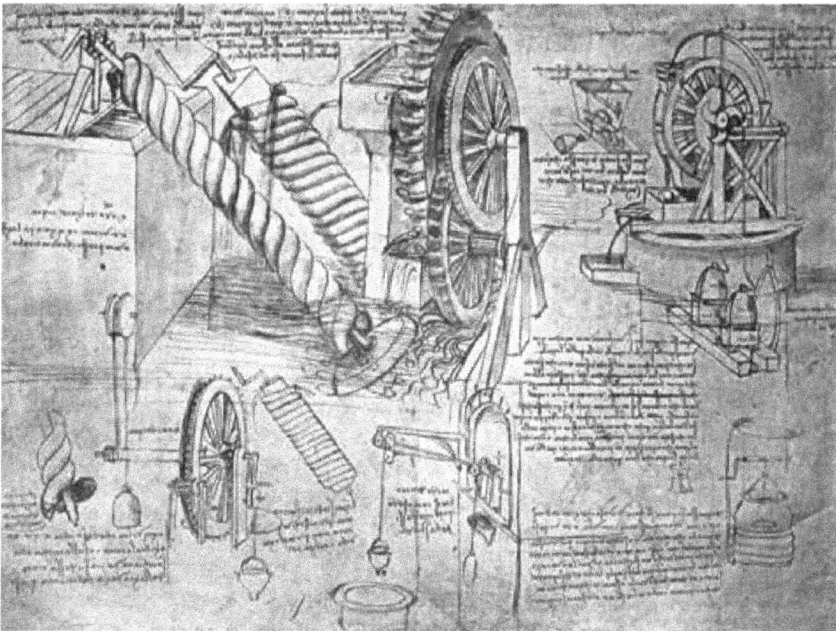

Figure 33.   Leonardo da Vinci (1452–1519), Codex Atlanticus. (Public Domain)

Since late Roman times, water wheels have also been used in mining operations (to transport water, ore, and personnel), in metallurgy, and for driving bellows. By the Industrial Revolution (17th century), waterwheels were already widely used in sawmills, textile and paper mills, and flour mills. Forges that used water-powered hammers and bellows were built along rivers or canals from which to derive power, and many different types of water wheels were used in the various workshops of the supply chain between raw materials and finished products. According to a survey commissioned by King Louis XV, there were about 140 integrated forges in operation in France in 1772, with many water wheels powering the different workshops of the production chain.

In 1765, the mechanical machine for making cotton cloth ("spinning jenny") was invented in England. It was a groundbreaking invention of the Industrial Revolution. The original hand-powered machine, which led to a high increase in productivity compared to the spinning wheel, was soon developed into more complex, machine-powered devices. Water-powered mills ("cotton mills") often served as prime movers, and they sprang up throughout England, France, and, later, the US.

Waterpower became increasingly important in the mining industry. In Germany, numerous dams and canals were built between the 16th and 18th centuries to provide waterpower for the many expanding mines. In the Harz Mountains, a low mountain range near Hanover, more than 100 dams were built with many canals, boasting a total length of more than 400 km. In the second half of the 19th century, the vertical water wheel was further developed and made even more powerful to satisfy the rapidly growing hunger for energy in industry. From the 1800s to the present, countless inventors and engineers have helped to further improve the performance of water turbines. Thanks to their advances, the modern Francis turbine today achieves over 95% efficiency.

In 1895, the large hydroelectric power plant at Niagara Falls ("Horseshoe Falls") was put into operation. The Edward Dean Adams power plant was originally equipped with five alternating current (AC) generators based on a design by Nikola Tesla, the inventor of the AC system. Each of the generators produced 5,000 horsepower, which was transmitted to New York City by an overhead transmission line. The Niagara Falls project launched the second phase of the Industrial Revolution and shaped the way that energy would be generated and delivered from then on — decentralized and in large quantities. The Hoover Dam in the US was built between 1931 and 1936 and dedicated on September 30, 1935 by President Franklin D. Roosevelt. After its modernization from 1986 to 1993, the Hoover Dam's 17 turbine generators have a total maximum capacity of 2,080 megawatts.

In the following decades, larger and larger dams with even greater turbine capacity were built around the world. In 1984, the Itaipu Hydroelectric Dam on the Parana River, located on the border between Brazil and Paraguay, was opened with an installed capacity of 14 GW (gigawatts). In 1994, the American Society of Civil Engineers selected Itaipu Dam as one of the Seven Wonders of the Modern World. As of 2021, the largest power plant ever built is the Three Gorges Dam in China, with a total capacity of 22,5 GW. This hydroelectric plant alone produces as much electricity as about 15 large, expensive, and dangerous nuclear power plants. Today, about 40% of renewable energy is generated from hydropower.

References

Buffalo Architecture and History. (n.d.). https://www.buffaloah.com/a/nf/adams/index.html

Center for Climate and Energy Solutions. (n.d.). Renewable energy. https://www.c2es.org/content/renewable-energy/

Encyclopaedia Britannica. (n.d.). Belisarius. https://www.britannica.com/biography/Belisarius

Hopley, C. (2023). A history of the British cotton industry. British Heritage. https://britishheritage.com/history/history-british-cotton-industry

International Energy Agency. (2021). Hydropower has a crucial role in accelerating clean energy transitions to achieve countries' climate ambitions securely. Press release, June 30. https://www.iea.org/news/hydropower-has-a-crucial-role-in-accelerating-clean-energy-transitions-to-achieve-countries-climate-ambitions-securely

Kloster Maulbronn. (n.d.). Water management. https://www.kloster-maulbronn.de/en/interesting-amusing/collections/water-management

Statistia. (n.d.). Largest hydroelectric dams worldwide as of 2021, based on power generation capacity. https://www.statista.com/statistics/474526/largest-hydro-power-facilities-in-the-world-by-generating-capacity/

The National Archives. (n.d.). Domesday: Britain's finest treasure. https://www.nationalarchives.gov.uk/domesday/

Wikipedia. (n.d.). Barbegal aqueduct and mills. https://en.wikipedia.org/wiki/Barbegal_aqueduct_and_mills

Wikipedia. (n.d.). Cistercians. https://en.wikipedia.org/wiki/Cistercians

Wikipedia. (n.d.). Hoover Dam. https://en.wikipedia.org/wiki/Hoover_Dam

Wikipedia. (n.d.). Ship mill. https://en.wikipedia.org/wiki/Ship_mill

Wikipedia. (n.d.). Water wheel. https://en.wikipedia.org/wiki/Water_wheel

Tidal Power

S ince time immemorial, man has tried to harness the power of the
tides. Tides are created by the orbital and gravitational interactions
of the Earth, sun, and moon. Large-scale, commercially successful
harnessing of tidal power is still in its infancy, and no technology has yet
harnessed the tides. Tidal power has major advantages over wind and solar
power — it is constantly available, predictable, and safe. Currently, however,
significant investments in research and development of this technology are
necessary to bring the cost down to a competitive level.

Tidal generators convert the energy of tidal currents into electricity. Higher
tidal current speeds dramatically increase electricity generation. The world's
largest tidal power plant is located at Lake Sihwa ("Sihwa-ho"), an artificial
lagoon in South Korea. Commissioned in 2012, the plant has a capacity
of more than 250 MW (megawatts) and a cost of more than $500 million.
The even larger MeyGen tidal power plant, with an estimated capacity of
398 MW, is currently under construction in the Pentland Firth in northern
Scotland. The first phase of the project began in April 2018; Phase 2 is due
to be commissioned in 2027. Once fully operational, MeyGen will be the
world's largest tidal energy plant, with an expected lifetime of 25 years.

The problem with tidal power plants is storage. The solution to this problem,
called Ocean Battery, was invented by Ocean Grazer, a Dutch start-up and
spin-off from the University of Groningen. The mechanism behind this
offshore battery system is based on dam technology and is a proven, reliable,
and efficient apparatus. The Ocean Grazer system pumps water from fixed
reservoirs into flexible bubbles on the seabed, where the energy is stored
under high pressure. When electricity is needed, the water flows from the

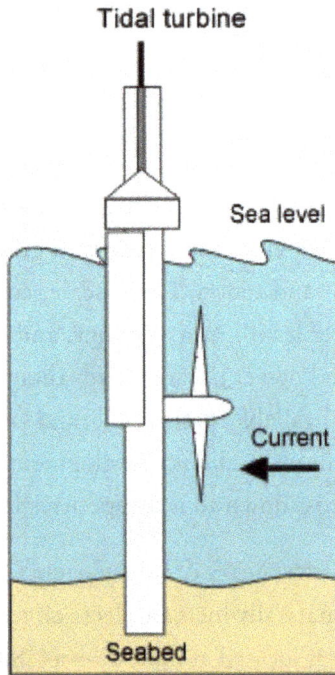

Figure 34.   Tidal turbine. (Public Domain)

flexible bubbles back into the fixed, low-pressure reservoirs and drives water turbines to generate electricity.

References

Tethys. (n.d.). MeyGen Tidal Energy Project — Phase I. https://tethys.pnnl.gov/project-sites/meygen-tidal-energy-project-phase-i

US Energy Information Administration. (n.d.). Hydropower explained. https://www.eia.gov/energyex-plained/hydropower/tidal-power.php

Wikipedia. (n.d.). List of tidal power stations. https://en.wikipedia.org/wiki/List_of_tidal_power_stations

Wikipedia. (n.d.). Sihwa Lake Tidal Power Station. https://en.wikipedia.org/wiki/Sihwa_Lake_Tidal_Power_Station

Geothermal Energy

G eothermal energy is generated from the Earth's internal heat. Geothermal power plants use hydrothermal resources, which contain both water ("hydro") and heat ("thermo"; in Greek, "thermos" means "warm"). These resources are tapped by drilling deep wells through which steam or hot water is brought to the surface. The hot water or steam drives a turbine that generates electricity. Some geothermal wells are up to three kilometers deep.

Figure 35. The Nesjavellir Geothermal Power Plant in Þingvellir, Iceland. (Public Domain)

Geothermal energy can also serve as an additional source of clean energy for electricity generation, but first and foremost, it is part of our civilization and history. Since ancient times, people have settled near geothermally active areas not only to enjoy the benefits of thermal water but also to cook or use volcanic products. The true "fathers of the geothermal industry" were the Etruscans, a civilization that can be traced back to Tuscany, Italy, since c. 890 BC. The Etruscans were pioneers not only in trade but also in craftsmanship. They were the first to coat their tools with enamel. To do this, they used borax, a boron compound found in boron-bearing springs that turns into insulating glass at high temperatures.

In the 18th century, thermodynamics, which entails the conversion of thermal energy to or from other forms of energy, made great strides. Scientists learned to convert steam into mechanical energy with increasing efficiency and into electricity using turbines and generators. The final step to geothermal energy was within reach and would not take long to be completed.

Tuscany is rich in thermal springs. In 1772, boric acid deposits were discovered there, in the vicinity of the village of Pomerance near Pisa. Borax (or sodium borate) was used for a long time, mainly as a flux in metalworking. Because it was expensive, it was used primarily by goldsmiths, silversmiths, and jewelers. Today, borax is an essential element for a wide range of applications in various industries as a cleaning agent, pesticide, and preservative, as well as in medicine and the production of glaze and enamel.

In 1818, French industrialist and engineer François Jacques de Larderel (1790–1858) pioneered the extraction of boric acid from the hot springs of Tuscany. He was the first to use geothermal energy to produce borax. In 1904, Piero Ginori Conti (1865–1939) built the first small geothermal power plant in Larderello. In 1913, the power plant was expanded to a capacity of 250 kW, and in 1916, two additional power plant units were installed, each with a capacity of 3.5 MW. Currently, the geothermal power plant in Larderello generates 545 MW of electricity.

After World War II, the US became the world's largest producer of geothermal energy. In the Mayacamas Mountains, about 100 kilometers north of San Francisco, naturally occurring hot water fields beneath the earth's surface are used to generate clean, green, renewable energy for homes and businesses throughout Northern California. The Geysers is the largest geothermal power complex in the world; it spans 45 square miles along the Sonoma-Lake County border. The operating company, Calpine, maintains 18 power plants at The Geysers with a capacity of about 725 MW — enough to power 725,000 homes, or a city the size of San Francisco.

As the climate crisis creates new opportunities and new needs, geothermal energy is returning to the international energy agenda. In 2022, UK-based deep geothermal energy company CeraPhi Energy Ltd announced plans to develop and implement geothermal projects to produce green hydrogen. The company is partnering with the financial services firm Climate Change Ventures. CeraPhi claims that geothermal energy is "the cleanest, cheapest, and most efficient energy for 24/7 baseload." The company is promoting a "geothermal revolution" with "geothermal everywhere." It announced that its innovative technologies will provide more than 500 GW (gigawatts) of additional energy over the next 30 years, representing about 7% of the world's growing renewable energy market.

References

CeraPhi. https://ceraphi.com/

Climate Change Venture. https://ccventures.io/

The Geysers. https://geysers.com/

Unwin, J. (2019). The oldest geothermal plant in the world. Power Technology, October 8. https://www.power-technology.com/features/oldest-geothermal-plant-larderello/

Wikipedia. (n.d.). Borax. https://en.wikipedia.org/wiki/Borax

Wikipedia. (n.d.). Geothermal energy. https://en.wikipedia.org/wiki/Geothermal_energy

Wikipedia. (n.d.). Geothermal energy in the United States. https://en.wikipedia.org/wiki/Geothermal_energy_in_the_United_States

Wikipedia. (n.d.). Larderello. https://en.wikipedia.org/wiki/Larderello

Wikipedia. (n.d.). The Geysers. https://en.wikipedia.org/wiki/The_Geysers

V. From Carbon to Hydrogen

Batteries and Fuels

T he big problem with renewables is not their generation but their storage. This is the key to the entire renewable energy process: We need to store the energy we generate so that we can use it when we need it. Today, there are various methods for this — batteries, pumped storage plants, and thermal and mechanical storage systems.

Technically, usable batteries have been around since the 19th century. Batteries convert stored chemical energy into electrical energy. Thanks to technological advances, large grid-scale battery systems are now experiencing record growth. Currently, large battery storage systems are being built in the United States (US), Germany, Australia, the United Kingdom (UK), Japan, South America, and many other countries. TesVolt is a German manufacturing company that has merged the two electric pioneers, Tesla and Volta, into its company name. The company offers a broad product portfolio combining state-of-the-art prismatic lithium battery cells with the patented Active Battery Optimizer (ABO) intelligent cell control system; it operates Europe's first-ever Gigafactory for battery storage systems. More than 2,000 plants with around 365 GWh of battery storage capacity have already been commissioned worldwide. The growth rate is extremely high, as the cost of design and construction has dropped rapidly. Nevertheless, the output and capacity of even the largest battery storage power plants are still an order of magnitude smaller than those of the largest pumped storage power plants.

Tesla, Inc. is known as the world's largest manufacturer of electric vehicles (EVs). The company also designs, sells, and installs solar energy and battery systems and other related products under the brand name Tesla Energy. The company has developed a home energy storage system called the

"Powerwall." Tesla has also acquired SolarCity, a manufacturer of solar panel systems, to add photovoltaic (PV) systems to its portfolio. Tesla Energy is currently building the largest battery storage systems in the world. One of them is located at Moss Landing in Monterey County, California, US. The facility was commissioned in 2021. It stores energy generated by the natural gas power plant during the day and releases it at night or when needed. The storage facility has a 300-megawatt lithium-ion battery consisting of 4,500 stacked battery racks.

The market situation is clear: Sales of batteries in the US will grow from $3.1 billion in 2020 to an estimated $15.04 billion in 2027. Tesla CEO Elon Musk is certain that "[i]n the long term, I expect Tesla Energy to be of the same or roughly the same size as Tesla's automotive sector or business. I mean, the energy business collectively is bigger than the automotive business."

References

Lambert, F. (2021). Tesla's new world's-largest battery is showing progress in drone flyover. Electrek, February 22. https://electrek.co/2021/02/22/tesla-new-world-largest-battery-progress-drone-flyover/

Tesla. https://www.tesla.com/en_sg/megapack

TesVolt. https://www.tesvolt.com/en/

Wikipedia. (n.d.). History of the battery. https://en.wikipedia.org/wiki/History_of_the_battery

Battery Materials

T he growth of the global battery business inevitably leads to further raw material depletion. New deposits of lithium, one of the most important components of battery storage systems, are already being developed all over the world. In Maine, at Plumbago Mountain in Newry (US), the world's largest known hard rock lithium deposit was found in October 2021. Almost at the same time, Europe's largest lithium deposit, with a capacity of 15 million tons of this "white gold," was discovered in Germany. From this, 40,000 tons of lithium hydroxide could be extracted annually. Depending on battery performance and battery type, this is enough to equip one million EVs.

Lithium is a soft, silvery to grayish-white (it turns yellow when exposed to air), odorless alkali metal. It is the least dense solid element and is both highly reactive and flammable. Although it is a metal, it is very light (it floats on water) and soft (it can be cut with a knife). Lithium was one of the first elements formed in the Big Bang, along with hydrogen and helium. It is very dangerous to humans; it can cause severe irritation and burns to the skin and eyes when inhaled and can irritate the nose, throat, and lungs. The environmental impact of lithium mining, which is necessary to replace fossil fuels with battery technologies, could become a major problem for the environment — and for the health of miners and workers.

South America, in the "Lithium Triangle" that covers parts of Argentina, Bolivia, and Chile, could be home to more than half of the world's lithium reserves. The ground beneath Bolivia's salt flats is believed to contain the world's largest single deposit of this valuable metal. To extract the lithium, miners first drill a hole into the salt flats. Then, the salty, mineral-rich brine is pumped to the surface. The brine is stored in large ponds for many months

Figure 36. Chemetall Foote Lithium Operation. (Source: Doc Searls, CC BY 2.0)

to allow any excess water to evaporate. To separate the lithium from other minerals in the brine, about 500,000 gallons (approx. 1,900,000 litres) of water are needed per ton of lithium. In the Salar de Atacama, Chile's largest salt desert, mining already consumes 65% of the region's water. Indigenous local farmers have started protesting against the extraction of lithium, saying that private interests are cashing in at the expense of the environment. For local farmers, the lithium rush in this region is a disaster.

Mining lithium not only requires a lot of water, but there is also the risk of toxic chemicals leaking from the evaporation ponds. These chemicals include hydrochloric acid and other hazardous waste products that are filtered out of the brine at each evaporation stage. In Tibet, BYD, the world's largest supplier of lithium-ion batteries, was forced to close a mine in 2013 after a leak of toxic chemicals from the lithium mine destroyed the local ecosystem.

Demand for lithium is rising dramatically as more and more EVs and electric storage batteries are produced worldwide. A Tesla "Model S" uses about 12 kilograms of the metal, while large storage systems obviously require much more. According to the consulting firm Cairn Energy Research Advisors, demand for lithium-ion batteries will increase tenfold over the next decade. Lithium mining, like all mining, is invasive, destroying the water table and polluting the soil and water. However, lithium is not the only problematic component of modern battery technology.

Cobalt and nickel are two other components of batteries that have enormous environmental and social costs when mined. Cobalt is a hard, gray metal found in rocks and soils. About 70% of the world's cobalt deposits are in the Democratic Republic of the Congo. Cobalt is extremely dangerous; it causes skin allergies and asthma-like allergies and can damage the heart, thyroid, liver, and kidneys. The New Jersey Department of Health has classified cobalt as a carcinogen, so it must be handled with extreme caution.

In 2020, the global production volume of cobalt was about 140,000 tons. The Democratic Republic of the Congo contributed a production volume of 95,000 tons. In the Congo, 15–30% of cobalt is mined in small-scale mining. Child labor, fatal accidents, and violent clashes between miners and security forces are commonplace. Given that small-scale mining is part of the supply chain and a lifeline for millions of the poorest Congolese, and the demand for cobalt is growing steadily and rapidly, the horrendous working conditions in the Congo will maintain to be tied to the global battery business. That is why it is important to continue highlighting the human rights situation of the more than 40 million miners around the world. They lead poorly paid and dangerous lives so the rest of the world can have a greener, more sustainable future.

Nickel is a completely different matter. The metal is the fifth most abundant element on Earth and was used by humans as early as 3500 BC. Nickel is ubiquitous in the biosphere; it is a hard, silvery-white metal that can cause skin irritation. Stainless steel is the largest end-use for nickel, accounting for two-thirds of total consumption. It is one of the most corrosion-resistant metals in the world and is also used in electroplating, in which a thin layer of nickel is applied to a metal object to beautify or protect it from corrosion and wear. Today, it is increasingly used for batteries. Nickel is a component of novel, bipolar nickel hydroxide batteries for EVs.

In 2020, global nickel production totaled 2.2 million tons. Indonesia is the country with the largest known nickel reserves, followed by Australia and Brazil. According to the US Geological Survey, total nickel reserves worldwide are estimated at 94 million tons. In 2020, the Russian company Norilsk Nickel was the world's largest nickel producer, followed by the Brazilian company Vale and the Swiss company Glencore, which also trade in numerous other minerals and metals.

Figure 37.   Norilsk. (Source: Ninaras, CC BY 4.0)

Nickel mining and smelting are associated with soil erosion, massive damage to terrestrial vegetation, and other ecological disasters, including a decline in the number and diversity of species and a reduction in biomass. Studies have demonstrated an increased risk of lung and nasal cancer in nickel refinery workers exposed to nickel dust. Women employed in nickel-contaminated work areas are at increased risk of spontaneous abortions in early pregnancy, and congenital heart defects may also be associated with exposure to nickel. Life expectancy for workers in Norilsk, Russia, the largest heavy metal smelting complex in the world, is 10 years lower than the Russian average. The industrial complex there also releases two million tons of sulfur dioxide annually, leading to an increased incidence of respiratory disease and cancer.

References

Bundesanstalt für Geowissenschaften und Rohstoffe. (2021). Lieferketten und Abbaubedingungen im artisanalen Kupfer-Kobalt-Sektor der Demokratischen Republik Kongo. April. https://www.bgr.bund.de/DE/Themen/Min_rohstoffe/Downloads/lieferketten_abbaubedingungen_artisanaler_Cu-Co-Sektor_DR_Kongo_de.pdf?__blob=publicationFile&v=3 (in German)

Cultural Survival. (n.d.). Isolated and impacted by nickel mining: Indigenous communities in Russia search for avenues of justice. https://www.culturalsurvival.org/publications/cultural-survival-quarterly/isolated-and-impacted-nickel-mining-indigenous-communities

GoContractor. (2018). Designing a nickel mining safety orientation. July 27. https://gocontractor.com/blog/nickel-mining-orientation/

Gross, T. (2023). How "modern-day slavery" in the Congo powers the rechargeable battery economy. National Public Radio, February 1. https://www.npr.org/sections/goatsandsoda/2023/02/01/1152893248/red-cobalt-congo-drc-mining-siddharth-kara

Harvard T.H. Chan School of Public Health. (2023). The dangers of cobalt mining in the Congo. February 16. https://www.hsph.harvard.edu/news/hsph-in-the-news/the-dangers-of-cobalt-mining-in-the-congo/

Institute for Rare Earths and Metals. (n.d.). Strategic metals. https://en.institut-seltene-erden.de/rare-earths-and-metals/strategic-metals-2/

ISS Insights. (2023). Nickel: Supply risks and ESG issues. April 12. https://insights.issgovernance.com/posts/nickel-supply-risks-and-esg-issues/

Nickel Institute. (n.d.). Nickel and the environment. https://nickelinstitute.org/en/policy/nickel-and-the-environment/

Statistia. (2022). Reserves of lithium worldwide as of 2022, by country. https://www.statista.com/statistics/268790/countries-with-the-largest-lithium-reserves-worldwide/

Tracy, B.S. (2022). Critical minerals in electric vehicle batteries. Congressional Research Service, August 29.

New Types of Energy Storage

Pumped storage hydroelectric plants (PSHs) are by far the most powerful form of energy storage. In this technology, water is pumped uphill during periods of low energy demand; when demand is high, the water is pumped back downhill, where water turbines recover the energy. With an installed capacity of over 180 GW (gigawatts), PSHs currently account for 95% of all large-scale energy storage worldwide. The efficiency of this type of storage is about 80%, and it is currently the most cost-effective way to store large amounts of energy. Hydropower will remain the world's largest source of renewable power generation for a long time to come and will play a crucial role in reducing carbon dioxide (CO_2) in the energy system.

Thermal energy storage (TES) is mainly used in buildings and industrial processes. Excess energy, e.g., waste heat, is stored to be used later for heating, power generation, or cooling. Thermal energy can be stored in liquids such as oil or water or in solid materials such as sand, salt, or rock. A simple and well-known example of thermal energy storage is water tanks in buildings. TES offers great flexibility by decoupling power generation and power consumption, even over extended periods of time.

Molten salt is a form of TES used in solar power plants (CSPs). In this process, molten salt is pumped through a solar collector, where it is heated to more than 550°C and then stored in a well-insulated tank. Other similar technologies use hot silicon or molten aluminum. Thermal energy (heat) can also be stored in rock caverns, a technology that is playing an increasing role in Finland's energy system, especially for space (domestic) heating.

In 2021, the Fraunhofer Society, a German research institute, found a new way to store heat. Thermochemical storage allows thermal energy generated in the summer to be conserved and used during the cold winter. Fraunhofer has found the ideal material for this — zeolites, minerals composed of the elements aluminum, oxygen, and silicon. The minerals come from both volcanic and sedimentary rock, and there are dozens of other artificial, synthetic zeolites. They store heat by removing the water that is stored within the mineral. When water vapor is added to the material (pellets), heat is released. This means that you could store the heat of summer in a zeolite tank and convert that stored energy back into heat (or another form of energy, like electricity) in the cold winter.

Globally, the total installed thermal energy storage capacity was 234 GWh in 2019. According to the International Renewable Energy Agency, this figure could triple to more than 800 GWh by 2030.

Mechanical energy storage systems use flywheels, compressed air, or other mechanical systems for storage. This type of technology has been around for a long time; a potter's wheel is a very early example of it. Today, mechanical energy storage systems are robust alternatives to battery storage.

Flywheel energy storage systems (FESSs) or rotational kinetic storage systems (RKSSs) store energy as kinetic energy in a high-speed rotor (flywheel) connected to a generator. The flywheel is accelerated, usually by electricity, to a very high speed and stores the energy as rotational energy. These flywheel systems are ideal for short-term, quick-response emergency power, such as the energy recovery flywheel system being built for use in Formula 1 race cars. A 500 KW flywheel storage system is being tested at TU Dresden that can "smooth" the feed-in of electricity generated by wind turbines for the power system. Further development "of a market and application-oriented flywheel technology for renewable energy producers is underway" (DEMIKS 2 research project, TU Dresden).

Compressed air energy storage (CAES) systems store energy in a manner similar to pumped storage power plants, but CAES systems use compressed air stored in underground caverns instead of water.

A new and highly effective storage solution comes from Energy Vault, a Switzerland-based company. Energy Vault promises to develop storage solutions that "use the fundamental principles of gravity and kinetic energy to store and transport energy." The company's concept for energy storage is simple and perhaps ingenious: Huge concrete blocks are hoisted and stacked by crane in a tower-like structure. This stores energy when it is available and not immediately consumed. When the kinetically stored energy is needed, the concrete blocks, pulled back down by gravity, drive generators, which thereby produce electricity. This is the same principle that has been used for decades in hydroelectric plants, where water is pumped into the upstream reservoir and released downstream when the power is needed. Energy Vault uses huge, heavy concrete blocks instead of water.

This new system has enormous advantages over hydropower plants. It can be built anywhere in the world at a reasonable cost and is even more efficient than pumped storage systems. The Swiss start-up has already built a working prototype in the Ticino municipality of Arbedo-Castione in southern Switzerland. This is the model for the emerging large-scale electricity storage industry. US investors have already partnered with the Swiss company to improve grid stability with this novel approach to energy storage. The low-cost simplicity of this out-of-the-box energy storage system opens a new path for more large-scale storage opportunities around the world.

The Clean Air Task Force says: "We need carbon-free electricity available 24 hours a day, seven days a week, 365 days a year, and zero-carbon fuels to power a global energy system two times its current size." The Climate Crisis Advisory Group (CCAG) was created in response to this emergency — "a new advisory group whose mission is to set out a critical pathway for urgent global action on climate change."

Figure 38.　Test device made by Energy Vault in Castione, Canton Ticino, Switzerland. (Source: Keimzelle, CC BY-SA 4.0, https://commons.wikimedia.org/w/index.php?curid=114689098)

References

Bauer, T. (2022). Overview of molten salt technology in the CSP sector state of the art and current research. Euchemsil 2022, 28th Euchem Conference on Molten Salts & Ionic Liquids, Patras, Greece. https://elib.dlr.de/189538/1/2022-06-07%20Bauer%20-%20Euchemsil.pdf

Celsius. (2020). Thermal energy storage. Celsius Wiki, August 17. https://celsiuscity.eu/thermal-energy-storage/

Clean Air Task Force. https://www.catf.us/

Climate Crisis Advisory Group. https://ccag.earth/

Energy Storage News. https://www.energy-storage.news/

Energy Vault. https://www.energyvault.com/

HELIOSCSP. (n.d.). Concentrated solar power tower: Use molten salt as an energy storage system. https://helioscsp.com/concentrated-solar-power-tower-use-molten-salt-as-an-energy-storage-system/

International Renewable Energy Agency. https://www.irena.org/

Office of Energy Efficiency & Renewable Energy. (n.d.). Pumped storage hydropower. https://www.energy.gov/eere/water/pumped-storage-hydropower

Polis Mobility. (2022). Focus on storage technologies. January 6. https://www.polis-mobility.com/magazine/articles/focus-on-storage-technologies.php

Southwest Research Institute. (n.d.). Mechanical energy storage research. https://www.swri.org/industry/advanced-power-systems/mechanical-energy-storage-research

Wikipedia. (n.d.). Compressed-air energy storage. https://en.wikipedia.org/wiki/Compressed-air_energy_storage

Wikipedia. (n.d.). Flywheel energy storage. https://en.wikipedia.org/wiki/Flywheel_energy_storage

Wikipedia. (n.d.). List of pumped-storage hydroelectric power stations. https://en.wikipedia.org/wiki/List_of_pumped-storage_hydroelectric_power_stations

Raw Materials

Standards of living … they're rising daily.

— Roxy Music, *In Every Dream Home A Heartache*

Over the past 50 years, global resource extraction — that is, everything extracted from the ground or otherwise, such as fossil fuels, metals, minerals, and biomass — has increased from 27 billion tons per year to nearly 80 billion tons in 2020. Given current trends, that number is likely to rise to 100 billion tons by 2030 and 167 billion tons by 2060 if we do nothing and continue pursuing a growth- and profit-driven business-as-usual scenario.

Some resources, especially precious metals and rare earth minerals, are already in short supply. In 2017, an EU raw materials study designated valuable resources such as lithium, magnesium, and cobalt as "critical raw materials" (CRM). A total of 27 out of 78 raw materials was deemed crucial to the economy and vital to all industries across all stages of the supply chain. A smartphone contains up to 50 of these critical raw materials, all of which contribute to its small size, light weight, and high functionality. CRMs are also irreplaceable in solar panels, wind turbines, electric vehicle batteries, and energy-efficient lighting.

The exploitation of natural resources cannot continue to increase year after year. If we miss the chance to stop this trend, we will soon no longer be able to continue all the production cycles that make our current lifestyle and consumption patterns possible. There is a simple number that sums up the truth: The per capita carbon footprint of those living in the industrialized Global North needs to be reduced by two-thirds (66%). Looking at the truly rich, the numbers are even starker: According to a 2020 OXFAM study, the per capita carbon footprint of the richest 10% of the world's population needs to be reduced by 90%, while the poorest 50% could double their footprint.

Behind the word "lifestyle" is the word "values," as our values determine our lifestyle and thus our carbon footprint. This means that, in order to reduce our carbon footprint, we have to move from the mentality of "bigger is better" to "small is beautiful." It all depends on how much we value everyone's well-being — not just our own but the well-being of every other human. It depends on how much we value our jobs, our homes, our friends, and our neighbors. We need a fundamental shift in values away from consumption and toward well-being. An enormous shift in consciousness is simultaneously needed, through which we must welcome future happiness and say goodbye to the destructive insensitivity that has taken possession of our minds and lives in recent decades.

The path from fossil fuels to renewable energy relies on technologies that, in many cases, are already known, proven, and in place. However, some transitional technologies will be needed to satisfy the increasing energy demands of the world's growing population. Whether nuclear energy will play a major role in this remains an open question. Given the hidden costs of nuclear energy — the immense impact of out-of-control nuclear power plants and the completely unresolved waste problem — the continuation of this technology seems rather unreasonable. Once again, (almost) everything is already in place. But the time for action is now.

References

European Commission. (n.d.). Critical raw materials. https://single-market-economy.ec.europa.eu/sectors/raw-materials/areas-specific-interest/critical-raw-materials_en

Intergovernmental Panel on Climate Change. (2022). The evidence is clear: the time for action is now. We can halve emissions by 2030. April 4. https://www.ipcc.ch/2022/04/04/ipcc-ar6-wgiii-pressrelease/

Open Access Government. (2022). US and EU responsible for 74% of global resource extraction. April 12. https://www.openaccessgovernment.org/usa-eu-resource-extraction-global-south-raw-materials-climate-change-co2-emission/133628/

Thunberg, G. (2019). "Our house is on fire": Greta Thunberg, 16, urges leaders to act on climate. *The Guardian*, January 25. https://www.theguardian.com/environment/2019/jan/25/our-house-is-on-fire-greta-thunberg16-urges-leaders-to-act-on-climate

United Nations. (n.d.). The climate crisis — a race we can win. https://www.un.org/en/un75/climate-crisis-race-we-can-win

Electric Vehicles

The invention of the internal combustion engine was the greatest crime ever committed in the history of mankind.

— Gustav Grob (1937–2018), ex-CEO,
International Clean Energy Corporation

After the lead-acid battery was invented in 1859 by the French physicist Gaston Planté (1834–1889), the British inventor and entrepreneur Thomas Parker built the first mass-produced electric car (e-car) in Wolverhampton in 1884. Soon, more and more e-cars were taking to the roads, and on April 29, 1899, Belgian racing driver Camille Jenatzy broke the 100 km/h (62 mph) speed barrier in his rocket-shaped vehicle "La Jamais Contente" ("The Never Contented").

Around 1900, interest in mobility increased significantly. In London, a fleet of electric cabs advertised their "humming" services, and in Berlin, electric delivery trucks were the most common vehicles in the streets besides horse-drawn carts and carriages. In the US, most automobiles were electric because they had so many advantages over the slower and more unreliable cars with internal combustion engines — they ran faster and smoother, did not make as much noise, and did not emit exhaust fumes. In New York, the New York Electric Vehicle Association was formed, which included the truck division of General Motors, which operated more than 2,000 trucks in the metropolitan area.

Figure 39. Illustration of "La Jamais Contente," the first automobile to reach 100 km/h in 1899. (Public Domain)

The development of the gasoline-powered automobile originated in Germany — with inventors such as Gottlieb Daimler, Carl Benz, and Rudolf Diesel — after many unsuccessful developments by tinkerers such as Francois Isaac de Rivaz of Switzerland. Carl Benz patented the world's first practical motor car, the three-wheeled "Motorwagen Nummer 1" ("Motor Car Number 1"), in 1886. Gasoline-powered cars improved over time (electric starter, muffler, etc.) and also became more reliable; they had a longer range than e-cars, and gasoline was now available at special filling stations rather than having to be purchased at drugstores or pharmacies. When Henry Ford began mass-producing cars with internal combustion engines (the "Model T" in 1912), the purchase price of gasoline-powered cars dropped dramatically, making e-cars increasingly unattractive. Ultimately, the oil industry won the war to power cars, trucks, delivery vehicles, and everything else on the road. Environmental aspects were irrelevant, and the automobile not only had practical use but also stepped into a significant role as a status symbol in society.

The development of the internal combustion engine made the automobile and many other means of travel and transportation possible. Tragically, the exhaust fumes from billions of combustion engines and those of the industries' refineries have brought the whole world to the edge of the abyss.

Today, our roads are dominated by roughly 1.5 billion cars, trucks, vans, and buses, most of which run on fossil fuels. We need to clean up this mess in less than a generation, as transportation accounts for about 14% of all greenhouse gas (GHG) emissions. To meet the cleanup target, these emissions must be reduced to zero by 2050. Thus, private transport and freight must be switched from gasoline engines to other forms of engines as quickly as possible, or the climate will irrevocably tip. Cars play a central role in the shift to other energy sources. Once the automotive switch to other energy sources has been made, other sectors of industry will follow suit.

References

EnergyBC. (n.d.). The history of the automobile. http://www.energybc.ca/cache/oil2/inventors.about.com/library/weekly/aacarsgasa6fc5.htm

Internet Archive. (n.d.). Electric vehicles. https://archive.org/stream/electricvehicles91916chic/electricvehicles91916chic_djvu.txt

Melosi, M.V. (n.d.). The automobile and the environment in american history. Automobile in American Life and Society. http://www.autolife.umd.umich.edu/Environment/E_Overview/E_Overview3.htm

Rebalance. (n.d.). Car as status symbol. https://rebalancemobility.eu/car-as-a-status-symbol-and-alternatives/

Wikipedia. (n.d.). History of the electric vehicle. https://en.wikipedia.org/wiki/History_of_the_electric_vehicle

Wikipedia. (n.d.). La_Jamais_Contente. https://en.wikipedia.org/wiki/La_Jamais_Contente

Modern Electric Vehicles

The current concept of mobility focuses on private motor vehicles, which are often occupied by only one person — the driver. For many, this is tremendously convenient and easy, and every year, there are new and faster models released to the market. However, this system is too expensive, too polluting, and too inefficient. All in all, it is very bad for the planet, so something has to change now. Fortunately, it is already happening.

In 2018, just over two million EVs were sold worldwide. Since then, sales have risen steeply, with more than ten million new EVs registered worldwide in 2022. This upward trend reflects people's realization that the switch to EVs is necessary to solve environmental problems related to cars, such as noise, particulate matter, and CO_2 emissions. In the long term, this switch will also enable a more flexible power grid, one based exclusively on renewable energy. On the other hand, more EVs on the roads will create an urgent need for a new charging infrastructure that is not currently available. Of course, more vehicles on the roads also means more traffic, more congestion, and an increasing depletion of the materials needed to build these new cars.

The problem of increasingly dense traffic can be solved in part by autonomous vehicle technology, which is developing at an ever-increasing pace. Autonomous EVs could guarantee reliable and ultimately cost-effective mobility for a rapidly evolving society. However, this development depends entirely on a secure 5G-based Internet. The 5G network is hundreds of times faster and has 1,000 times the capacity of the current 4G network. The new 5G system will make the machine-to-machine communication required for autonomous driving possible and will also be able to provide improved broadband connectivity for cell phones, as well as super-fast and resilient data transmission on the Internet.

On the political side, a whole range of supporting measures is needed to accelerate this process. For example, it is essential to promote the sale of EVs through subsidies and to expand the network of charging stations. These incentives will accelerate the change and help replace the advantages of conventional, gasoline-powered vehicles, such as the existing, dense service station infrastructure, with new structures.

The most critical part of EVs is their battery. However, with today's rapid advances in battery technology, it is foreseeable that higher battery capacities and falling battery manufacturing costs will mean that EVs will be able to compete with internal combustion vehicles in terms of price within the next decade. In addition, mobility services like Ola, Lyft, Uber, and Didi are already pointing the way to a shared mobility solution via their e-fleets.

Over the past decade, the EV market has spawned new automakers and brought about a late but rapid change in the portfolios of traditional automakers. It seems that around 2030, there will be few new cars with internal combustion engines left on the market. Tesla, Lucid, Polestar, and Porsche are automakers at the top end of the passenger car market, while Volvo, Daimler-Benz, and others are pushing their production of alternatively powered cargo vehicles. In this sector, hydrogen will ultimately win the race as the cleanest fuel while also being cheap. In the low-cost passenger car market, only Chinese manufacturers like BYD have a longer list of affordable (or maybe even "cheap") EVs. This gap needs to be closed because in a world struggling to survive, it does not make sense to drive around — or be driven around — in electric, hybrid, or hydrogen-powered cars costing $100,000 and up.

As far as electrification is concerned, the e-bus market has developed the quickest. With diesel-powered buses being one of the main causes of increasing pollution in already congested urban centers, the move to electrified buses (e-buses) was long overdue. In 2003, Wang Chuanfu, CEO of BYD, the Chinese battery manufacturer, established a subsidiary for automotive production. In Shenzhen, a city of 13 million, a fleet of 16,000 e-buses and 22,000 e-taxis now helps combat the city's nightmarish air pollution. BYD has been able to grow so quickly because it has received more than $1 billion in government grants and subsidies. This is part of the Chinese government's $50 billion pledge to become a world leader in EV production. Today, BYD already produces its buses in California and promises to build up to 1,500 e-buses per year.

Proterra is an e-bus manufacturer in the US. The company has made a similar leap forward as BYD but with market investment rather than government incentives. After 10 years of research and development, the company has won several tenders for municipal school bus transportation in the US, with a new vehicle platform and battery technology. Proterra has already partnered with the world's largest commercial vehicle manufacturer, Daimler-Benz, and promotes its product portfolio as a "clean, quiet transportation solution for all."

Another problem in EV technology is battery recycling because at the mechanical end of an EV's life of about 10–12 years, the battery may still be relatively good. Jeffrey Brian "JB" Straubel, a Stanford graduate, was the chief technology officer of Elon Musk's visionary automotive company, Tesla Motors, for 17 years. Now, Straubel wants to turn lithium-ion battery manufacturing into a circular economy, not just for EV batteries but also for batteries used in cell phones, tablets, and laptops.

Founded in 2019 by Straubel, Nevada-based Redwood Materials is working on nothing less than revolutionizing battery recycling. In an interview,

Straubel explained his intentions, stating, "Assuming a future where virtually all transportation logistics will be fully electrified, we'll hardly need to mine any new raw materials." Straubel expects the recycling rate of metals used in lithium-ion batteries to reach 95–98%. The range of raw materials that can be extracted and recycled from EV batteries and other batteries is wide — lithium, nickel, copper, gold, silver, cobalt, palladium, tantalum, neodymium, and many other valuable, critical raw materials. The interest in this new recycling technology is huge; the car manufacturer Ford has already partnered with Redwood Materials. Redwood will build its first large recycling plant in the US. It is expected to be capable of supplying half of the EV battery production planned for North America by 2030.

The electrification of aging, diesel-powered bus fleets has already become a major goal for cities worldwide. For 2021, several countries — from the Netherlands to New Zealand — have set ambitious targets for urban electrification. In Latin America, the ZEBRA Bus Initiative has committed an additional $1 billion in investment to expand e-bus fleets across the region. In many countries, there are large public contracts for transportation electrification that were previously unimaginable. It has become clear to governments, businesses, and investors around the world that concrete action must be taken to implement the ambitious goals of the 2015 Paris Agreement — first and foremost for cities and public transport.

References

Berlin, A., Zhang, X., and Chen, Y. (2020). Case study: Electric buses in Shenzhen, China. https://iea.blob.core.windows.net/assets/db408b53-276c-47d6-8b05-52e53b1208e1/e-bus-case-study-Shenzhen.pdf

InsideEVs. https://insideevs.com/

Maurer, M. Gerdes, J.C., Lenz, B., and Winner, H. (2015). Autonomes Fahren: Technische, rechtliche und gesellschaftliche Aspekte. https://library.oapen.org/bitstream/id/05b73ab3-c816-4d0a-ba45-135140828f78/1002192.pdf (in German)

Proterra. https://www.proterra.com/

Redwood Materials. https://www.redwoodmaterials.com/

Royal Society of Chemistry. (n.d.). Policy, evidence and campaigns. https://www.rsc.org/policy-evidence-campaigns/brought-to-you-by-chemistry-podcast/

Sustainable Buses. (2023). Electric bus, main fleets and projects around the world. September 14. https://www.sustainable-bus.com/electric-bus/electric-bus-public-transport-main-fleets-projects-around-world/

Wikipedia. (n.d.). Redwood Materials, Inc. https://en.wikipedia.org/wiki/Redwood_Materials,_Inc

Hydrogen-Powered Vehicles

Hydrogen has the potential to address important challenges on the way to a net-zero world. It can help decarbonize sectors where electrification alone is not enough, such as steel or cement production. Hydrogen can also be used as a medium for intermediate energy storage, and it could be used as a fuel for fuel cell vehicles, aircraft, or ocean-going vessels.

Fuel cell electric vehicles (FCEVs) produce no exhaust emissions other than water vapor and heat. FCEVs use energy stored in the form of hydrogen, which is then converted into electricity. The first commercially produced hydrogen fuel cell car was introduced in 2013. As of December 31, 2020, some 16,000 FCEVs have been sold worldwide, a figure that clearly reflects market acceptance. "The hydrogen fuel cell dream is all but dead for the passenger car market" (The Motley Fool). Another factor is price; the hydrogen needed to move an FCEV one kilometer still costs about eight times as much as the electricity needed to move a battery-powered EV that same distance.

Many automakers that developed hydrogen cars have already switched to producing battery-powered EVs. Only three automakers have not yet abandoned fuel cells for passenger cars — Toyota, Hyundai, and BMW. Even so, hydrogen fuel cells currently appear to be the technology of choice only for some hard-to-electrify transportation segments, such as marine and aviation.

So, what is behind the hydrogen hype? One answer is as simple as it is perplexing: Most hydrogen gas used today is "gray," meaning it is produced from natural gas, a high-emission fossil fuel. The oil industry still promotes this byproduct of oil production, in which gray hydrogen is obtained through a process called "steam reforming." Green hydrogen, on the other hand, is the only kind of hydrogen that is produced in a carbon-neutral way. Only "green" hydrogen will be able to play an important role in the global effort to reduce net emissions to zero by 2050. However, if sufficient hydrogen can be made available to the world market quickly, i.e., within the next 8–10 years, this new technology could become the backbone of the entire world energy system.

Hydrogen-powered fuel cell trucks are currently only manufactured by Hyundai (South Korea) and Hyzon (US). Hyzon's heavy-duty trucks have fuel cells with the highest power density in the world. Hyundai has already delivered a road-going fleet of 10 hydrogen-powered trucks to Switzerland for use as delivery trucks. Diesel-engine trucks have the major problem of emitting a lot of CO2, so the future of the trucking segment will most likely hinge on the use of hydrogen. Thus, the Hyundai trucks could turn out to be harbingers of the future.

To transport hydrogen in large quantities, the shipping industry has already unveiled a suitable means of transportation — the *Suiso Frontier*. It is the first ship — its name "Suiso" means hydrogen in Japanese — that can transport hydrogen from its production site (Australia) to an international market (Japan). It is part of the Hydrogen Energy Supply Chain (HESC) project, which aims to demonstrate the feasibility of a continuous hydrogen supply chain between two countries. Unfortunately, the hydrogen on board the *Suiso* is produced from coal, so it is brown hydrogen, not green hydrogen. However, at least this experiment shows that hydrogen can be transported in large quantities and over long distances by ship. In the future, this will be the path of progress into a hydrogen-based energy economy, surely against the inevitable resistance of the oil industry, which still wants to produce and sell hydrogen produced from crude oil and natural gas.

References

Baxter, T. (2020). Hydrogen cars won't overtake electric vehicles because they're hampered by the laws of science. The Conversation, June 3. https://theconversation.com/hydrogen-cars-wont-overtake-electric-vehicles-because-theyre-hampered-by-the-laws-of-science-139899

Bigo, A. (2022). Hydrogen in transport: Everything you need to know in 10 questions. Polytechnique Insights, November. https://www.polytechnique-insights.com/en/columns/energy/hydrogen-in-transport-everything-to-know-in-10-questions/

Heid, B., Martens, C., and Wilthaner, M. (2022). Unlocking hydrogen's power for long-haul freight transport. McKinsey & Company, August 2. https://www.mckinsey.com/capabilities/operations/our-insights/global-infrastructure-initiative/voices/unlocking-hydrogens-power-for-long-haul-freight-transport

Hoium, T. (2019). Hydrogen cars appear dead as EVs take the reins. The Motley Fool, April 23. https://www.fool.com/investing/2019/04/23/hydrogen-cars-appear-dead-as-evs-take-the-reins.aspx

Hydrogen Council. (2022). Toward a new era of hydrogen energy: Suiso Frontier built by Japan's Kawasaki Heavy Industries. October 4. https://hydrogencouncil.com/en/toward-a-new-era-of-hydrogen-energy-suiso-frontier-built-by-japans-kawasaki-heavy-industries/

Hyundai Hydrogen Mobility. https://hyundai-hm.com/

Hyzon Motors. https://www.hyzonmotors.com/

US Department of Energy. (n.d.). Fuel cell electric vehicles. https://afdc.energy.gov/vehicles/fuel_cell.html

US Environmental Protection Agency. (n.d.). Hydrogen in transportation. https://www.epa.gov/greenvehicles/hydrogen-transportation

How to Fly Net Zero?

ydrogen is a great fuel because it has a high inertial energy content per unit of weight, is lightweight, has the best thrust-to-weight ratio of any fuel, and burns with extreme intensity (5,500°F). Through the Fly Net Zero initiative, International Air Transport Association (IATA) member airlines are committing to flying carbon-free by 2050. Implementing this pledge would bring high-carbon aviation in line with the goals of the 2015 Paris Agreement.

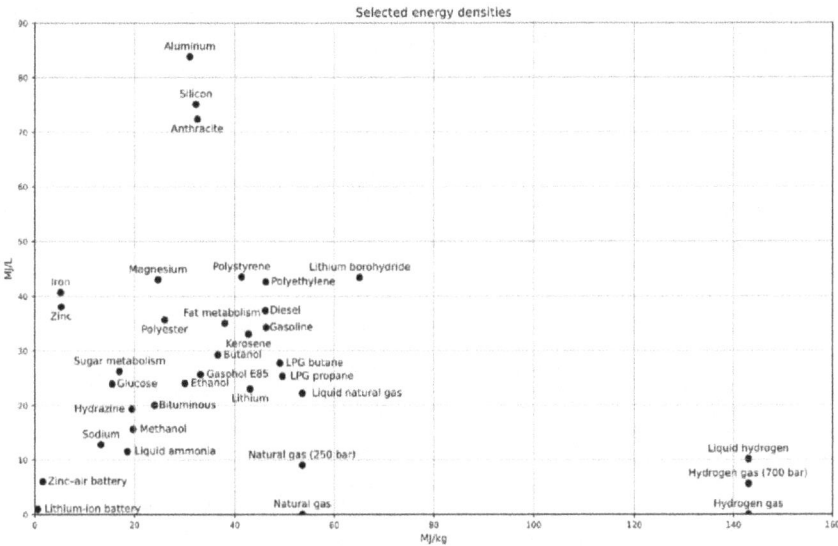

Figure 40.   Selected energy densities plot. (Source: Scott Dial, Public Domain)

Decarbonizing the aviation industry is indeed a major challenge. Projections indicate that more than 10 billion travelers will take a flight in 2050, about five times the number in 2021. Ten billion travelers translate into

1.8 gigatons of CO2 — or 2.5% of total CO2 emissions — that must be reduced. Under IATA's plans, 65% of the targeted reduction will come from sustainable aviation fuel (SAF), 19% from carbon capture, and the rest from "new technologies" and infrastructure improvements.

Airbus is the world's largest aircraft manufacturer. Its goal is to develop the world's first zero-emission commercial aircraft by 2035. The ZEROe project envisages liquid hydrogen as the basic fuel for the future. To turn this concept into reality, Airbus has set up a new innovation center (UpNext) and a Zero Emission Development Centre (ZEDC) for hydrogen technologies.

An emission-free, hydrogen-powered aircraft is already in the air. After ten years of research, development, and testing, German start-up H2FLY has unveiled the HY4 aircraft, which implements the concept of hydrogen-powered aircraft. The startup touts a "long-range, zero-emission, low-noise propulsion system." Hydrogen holds enormous potential for the future of air mobility; small air cabs, business jets, and regional jets could all be powered by clean, green hydrogen, but only if green hydrogen production is ramped up quickly.

Hypersonic Launch Systems, an Australian company, is taking a different approach. Its SPARTAN engine is based on the scramjet principle — a propulsion system ideally suited for supersonic flight. It is reusable, runs on hydrogen, and generates enough thrust to reach speeds of Mach 5 to Mach 12. This engine can be used for satellite launch systems and opens up new applications for hypersonic technologies.

References

Airbus. (n.d.). ZEROe: Towards the world's first hydrogen-powered commercial aircraft. https://www.airbus.com/en/innovation/zero-emission/hydrogen/zeroe

Hypersonix Launch Systems. https://hypersonix.com/

H2Fly. (n.d.). Breaking the hydrogen barrier. https://www.h2fly.de/

International Air Transport Association. (n.d.). Our commitment to fly net zero by 2050. https://www.iata.org/en/programs/environment/flynetzero/

International Energy Agency. (n.d.). Aviation. https://www.iea.org/energy-system/transport/aviation

Ritchie, H. (2020). Climate change and flying: What share of global CO2 emissions come from aviation? Our World in Data, October 22. https://ourworldindata.org/co2-emissions-from-aviation

Statista. (n.d.). Number of flights performed by the global airline industry from 2004 to 2021, with forecasts until 2023. https://www.statista.com/statistics/564769/airline-industry-number-of-flights/

The Methane Problem

M ethane is a combustible, colorless, and odorless gas with the chemical formula CH4. As the simplest of the hydrocarbons, it is part of the Earth's carbon cycle. It is produced during the conversion of organic matter in the absence of oxygen in swamps, bogs, thawing permafrost, or during the decomposition of algae on the ocean floor. Volcanic activity can also release methane. Depending on the deposit, natural gas consists of up to 98% methane. Other components of natural gas may include ethane, propane, and butane, the next higher hydrocarbons, as well as the non-combustible gasses carbon dioxide and helium. Like petroleum and coal, the methane in natural gas is formed by the conversion of biomass in the ground, so it is itself a fossil fuel. It can also naturally escape from its deposits and enter the atmosphere. However, there is also anthropogenic methane from rice cultivation, animal husbandry, the landfilling of waste, wastewater treatment, and, last but not least, the use of fossil fuels. Overall, 60% of atmospheric methane is of anthropogenic origin.

In the atmosphere, methane is oxidized to carbon dioxide (and water) by atmospheric oxygen. In the troposphere, the lowest atmospheric layer up to a height of about 10–18 kilometers, this oxidation happens very slowly and with the involvement of photochemical processes, so that a released methane molecule survives there for an average of somewhat less than ten years. During this time, however, it acts as a very powerful GHG that is much more potent than carbon dioxide. Over a period of one hundred years — a period commonly used for Global Warming Potential (GWP) — methane heats the earth 28 times faster than carbon dioxide, and in 20 years, it heats the earth 86 times faster. In fact, methane contributes about 20% of global GHG emissions, making it a relatively large contributor to global warming.

Methane Sources

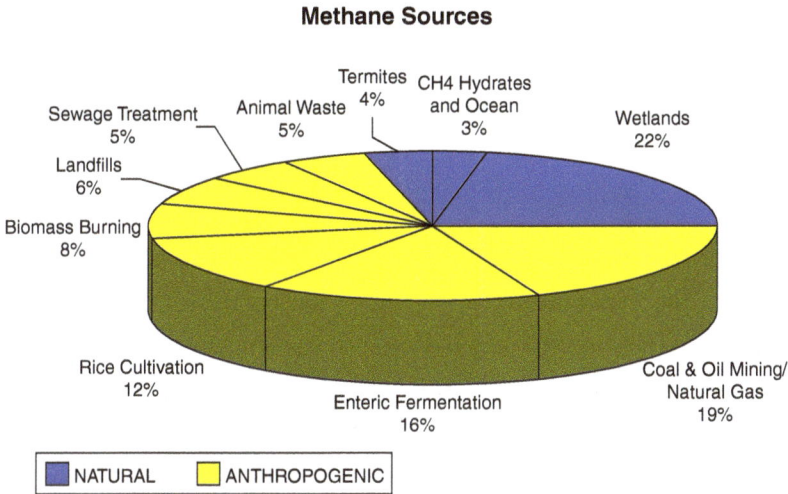

Figure 41.   Natural and anthropogenic methane sources. (Public Domain)

Compared to pre-industrial times, the methane content of the atmosphere has roughly tripled. Science has several explanations for this. One is oil production, but others are the increase in emissions from fracking, the warming of tropical or arctic swamps, and the increase in livestock farming.

According to the International Energy Agency (IEA), the oil and gas industry emitted around 70 million metric tons of methane in 2020. Seventy percent of these emissions escape through leaks at fossil fuel extraction facilities or during transportation. To stop these emissions, operators need to repair their plants and pipelines. New technologies are helping in this effort. In December 2019, for example, satellites from the European Space Agency (ESA) used high-resolution infrared cameras to detect a large methane leak in North America's Permian Basin, Texas.

Methane being released from permafrost soil is another matter. Permafrost covers 65% of Russia's land mass, and it is melting faster and faster. Rising temperatures in the far north of Russia (also in Canada and the US) are causing the release of huge amounts of methane. Methane is produced not only by the microbial decay of organic matter in the thawing permafrost — known as "microbial methane" — but is also released from natural gas trapped beneath or within the permafrost layer during thawing. This is referred to as "thermogenic methane."

The recent heat wave in 2020 was responsible for a further increase in emissions. The temperature rise in the region north of the Arctic Circle was almost 11°C above normal. This released ancient methane deposits trapped in the area. A study led by the University of Bonn, Germany, interpreted the data from its investigation as alarming, saying it could even "have dramatic impacts on the already overheated global climate." The scientists concluded that methane released from permafrost soils contributes more to climate change than previous studies had suggested. Nikolaus Froitzheim, lead author of the study, interpreted that the data "may make the difference between catastrophe and apocalypse."

The ongoing thawing of permafrost soils will release significant amounts of methane into the atmosphere in the near future. This is one of the reasons why we must do everything we can to stop global warming as soon as possible. The methane problem may not be the only surprise to appear in the midst of this climate crisis. We may have to prepare for several possible worst-case scenarios without knowing precisely where, when, or how they will occur.

ARPA-e is the US Advanced Research Projects Agency's program to advance cutting-edge energy technologies before they reach the private market. The agency focuses on transformative energy projects that develop entirely new ways to generate, store, and use energy. In 2021, ARPA-e announced funding for the Reducing Emissions of Methane Every Day of the Year (REMEDY) program. This three-year, $35 million research program is designed to reduce methane emissions from the coal, oil, and gas sectors.

This is a relatively small but absolutely necessary step in finding ways to reduce methane emissions as long as fossil fuels are being promoted, hopefully with the side effect of developing more technologies to reduce methane emissions in other sectors.

There is already another ARPA-e program (HESTIA) with a similar goal. The agency has funded a $45 million program to develop technologies that make buildings net carbon stores. This program could contribute to the needed revolution in construction that would make the construction of buildings a net carbon negative and the finished building itself a carbon storage facility.

References

Advanced Research Projects Agency — Energy (n.d.). Harnessing emissions into structures taking inputs from the atmosphere. https://www.arpa-e.energy.gov/technologies/programs/hestia

Advanced Research Projects Agency — Energy (n.d.). Reducing emissions of methane every day of the year. https://www.arpa-e.energy.gov/technologies/programs/remedy

Climate Reanalyzer. https://climatereanalyzer.org/

Froitzheim, N., Majka, J., and Zastrozhnov, D. (2021). Methane release from carbonate rock formations in the Siberian permafrost area during and after the 2020 heat wave. *Proceedings of the National Academy of Sciences*, **118**(32): e2107632118. https://www.pnas.org/doi/10.1073/pnas.2107632118

Kindy, D. (2021). Permafrost thaw in Siberia creates a ticking "methane bomb" of greenhouse gases, scientists warn. *Smithsonian Magazine*, August 5. https://www.smithsonianmag.com/smart-news/ticking-timebomb-siberia-thawing-permafrost-releases-more-methane-180978381/

What Color Is Hydrogen?

S teel production today almost exclusively uses fossil fuels. It will not be easy to replace the natural gas or coal needed for this production process with green hydrogen or electricity. However, since a rapid and bold change of direction is inevitable in many industrial processes, such as steel production, hydrogen will probably be the only sensible solution in the long term.

Hydrogen is an invisible gas. On Earth, it is found in almost unlimited quantities. However, hydrogen is only present in its chemically bound form, e.g., as water (H_2O), in various hydrocarbons (petroleum, natural gas, coal, biomass, etc.), or in other organic components. Therefore, pure hydrogen must first be produced in a controlled process using (a lot of) energy. Hydrogen is thus a carbon-free secondary energy carrier comparable to electricity or heat.

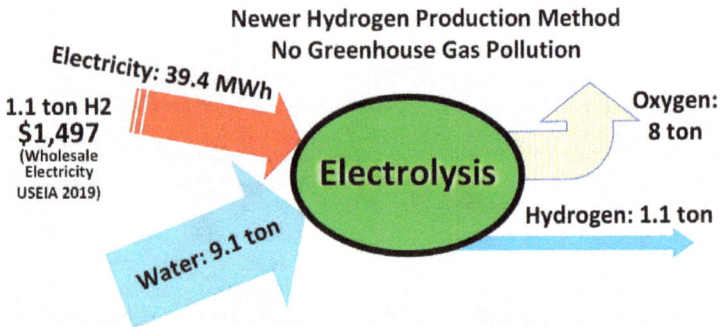

Figure 42.   Graphic of an industrial process showing inputs into electrolysis to produce one ton of hydrogen and other outputs. (Public Domain)

Hydrogen comes in many (color) varieties: There is green, gray, blue, brown, yellow, turquoise, and pink hydrogen. Green hydrogen is produced using clean electricity from renewable energy sources, such as solar or wind power, through the electrolysis of water. As the price of energy from renewable sources decreases, the price of green hydrogen will also decrease over time.

Blue hydrogen is produced by the steam reforming of natural gas in so-called steam methane reforming (SMR) or autothermal reforming (ATR) plants. This produces carbon dioxide (CO2) and hydrogen (H2), which are separated in a second process. Part of the CO2 may be used or stored in decommissioned natural gas reservoirs (by the Carbon Capture and Storage (CCS) process). Due to the amounts of CO2 and methane released during the extraction and transport of this natural gas, blue hydrogen has a relatively large GHG footprint.

Brown hydrogen is produced from lignite, while black hydrogen is derived from hard coal. The manufacturing process is called gasification, which is a rather complex process. It essentially converts coal into H2 and CO2, which escapes into the atmosphere. These two carbon-rich hydrogens account for about 95% of the current world production.

Turquoise hydrogen is produced by breaking down methane (natural gas) with solid carbon as a byproduct. This process is known as "methane pyrolysis" and is driven by heat generated with electricity. The resulting carbon is in solid form and can be used, for example, as a soil conditioner or to make tires. If the electricity used comes from renewable energies, the process is carbon-free.

Gray hydrogen is also produced from natural gas (methane), but in this case, it is made by the SMR process without GHG capture. This is a very high-carbon process that produces about 800 million tons of CO2 emissions each year. If some of the carbon is captured using the CCS method, which is costly and requires significant resources, the gray hydrogen is classified as blue hydrogen.

Yellow hydrogen is a relatively new term for H2 produced by electrolysis using solar energy. This promising form of H2 production is being driven by the Swiss start-up Synhelion. The company is a spin-off of an ETH Zurich research group led by Professor Aldo Steinfeld. For their EU project "Sun-to-Liquid," the research group and its partners received the Energy Globe World Award 2021 at the COP26 climate summit in Glasgow.

Pink (aka red) hydrogen is produced by electrolysis using nuclear energy. There are also rumors of white hydrogen, which may be obtained from beneath the ground. The most important point in the whole hydrogen color scheme is that H2 can only achieve the goal of being a clean energy source if there are no GHG emissions in its production.

Green hydrogen alone cannot stop climate change. It is still too expensive, as far too little green electricity is available to produce it. So-called turquoise hydrogen could also help. However, the energy required to produce the turquoise variant is high, and the climate would only be protected if the energy required could be made available without emissions.

In eastern Germany, in the Lausitz region, intensive work has been going on for years to convert the coal industry there to a green hydrogen industrial area. This is intended to facilitate the early phase-out of the coal industry and move further toward Germany's ambitious climate targets. The project focuses on the production of H2 systems (e.g., fuel cells, electrolysis, storage), power generation from H2 (centralized and decentralized), and H2-based mobility. The so-called "Gigawatt Factory" already had its groundbreaking ceremony, welcomed by Robert Habeck, the German Federal Minister for Economic Affairs and Climate Action, and is hailed as "important to the future of industry in Germany."

A large number of projects and initiatives in the growing H2 sector are already operating in the region: Reference Power Plant Lausitz, Reallabor

Lausitz, Fraunhofer Hydrogen Lab Görlitz, H2 and Storage Research Center, Innovation Campus Siemens Görlitz, and WALEMO Model Region. The advantages of locally producing green hydrogen are obvious; short distances mean local production and supply of the hydrogen, while the already existing gas network can easily be converted to locally produced green hydrogen.

In Germany, the share of privately used fuel cell vehicles was about 0.001% in 2019. Especially for commercial vehicles (trucks, buses), the demand forecast looks quite different. Government procurement of new vehicles, for example, envisages a 65% share of zero-emission vehicles as early as 2030. Fuel cell trucks and buses will account for a large portion of that total. This is an example of how targeted legislation can practically influence positive development toward zero-emission passenger and freight transportation.

Transporting H2 could also be a problem, but with the commissioning of the H2 cargo ship *Suiso Frontier*, it seems to have been solved, at least for transcontinental transport. Also, the world's first zero-emission, H2-powered bulk carrier has already entered service. The ship will transport grain from eastern Norway to western Norway and, on the way back, will ship rock or gravel. The cargo ship will be powered by emission-free green hydrogen in combination with rotor sails. This is an important milestone in establishing green hydrogen as an energy carrier for sea transport.

The use of H2 in the application of fuel cells for marine transportation is crucial for the whole shipping industry. This industry is responsible for a significant portion of global GHG emissions, and reducing its carbon footprint has become a pressing issue. Most vessels crossing the ocean still run on fossil fuels, and a huge number of these run on Heavy Fuel Oil. Its common use in ship engines is very harmful to the atmosphere as it has the highest level of exhaust emissions among marine fuels.

The use of green hydrogen in steel production will result in the huge transformation of an entire industry. Almost all known conventional steel-making plants will disappear, namely, blast furnaces, sinter processors, and coke plants. They will be replaced by fluidized bed reduction furnaces, so-called melting gasifiers, and electric furnaces. Instead of CO_2, the only byproduct of the fossil-free process is water. However, despite all the technical problems and new requirements, there is already a pilot plant for H2-based steelmaking, the HYBRIT project of steelmaker SSAB, located in Luleå, Sweden. The plant intends to bring fossil-free steel to the market as a commercial product in larger quantities in 2026.

References

Benham, H. (2023). Energy Transition's clean technologies are empowering an industrial policy revolution. Carbon Tracker, February 27. https://carbontracker.org/the-energy-transitions-clean-technologies-are-empowering-an-industrial-policy-revolution/

Castelvecchi, D. (2022). How the hydrogen revolution can help save the planet — and how it can't. *Nature*, November 16. https://www.nature.com/articles/d41586-022-03699-0

Green Steel World. (2022). The future of steel — hydrogen-based steelmaking. February 17. https://greensteelworld.com/featured-article-the-future-of-steel-hydrogen-based-steelmaking

International Energy Agency. (n.d.). Hydrogen. https://www.iea.org/energy-system/low-emission-fuels/hydrogen

Lloyd's Register. (2022). Norwegian zero-emission bulk carrier project awarded LR AiP. March 1. https://www.lr.org/en/about-us/press-listing/press-release/norwegian-zero-emission-bulk-carrier-project-awarded-lr-aip/

National Grid. (n.d.). The hydrogen colour spectrum. https://www.nationalgrid.com/stories/energy-explained/hydrogen-colour-spectrum

SSAB AB. https://www.ssab.com/de-de

Steiner, A. (2021). The energy revolution has arrived — here's how to be a part of it. United Nations Development Programme, July 28. https://www.undp.org/blog/energy-revolution-has-arrived-heres-how-be-part-it

Wikipedia. (n.d.). Hydrogen production. https://en.wikipedia.org/wiki/Hydrogen_production

Biofuels

An environmentally friendly energy transition requires billions of tons of biomass every year for the production of biofuels. Today, crops such as corn or soybeans are primarily used for the production of biofuels. However, growing these crops also has a negative impact on the environment: a lot of land is needed, and a choice must be made between using the crop as food or fuel. Shifting the focus from land to water — three-quarters of the earth is covered by water, after all — an alternative can be found — seaweed.

In 2019, San Diego, California-based Marine Bioenergy, Inc. and researchers at the USC Wrigley Institute for Environmental Studies began testing a revolutionary concept: growing kelp in the open ocean. The goal is to develop a system of floating kelp farms that will be scattered along the entire Pacific coast of the US. The system moves the kelp back and forth between nutrient-rich deep water and sunlight at the surface, and it can also be fully submerged to avoid storms and ships. The open ocean is a vast, untapped area for biomass production, and this novel concept makes it possible to grow giant kelp even in the vast regions of our blue planet.

This kelp concept is not in competition with food production on agricultural land, as kelp farms can be established anywhere in the open ocean. Kelp is one of the fastest-growing marine plants on earth (a "protist"), with growth rates of up to 30 centimeters per day. It can easily be processed into fuel and continues to grow throughout the year, providing a constant source of biomass. The idea is to produce millions of tons of kelp-based biomass to make biofuels at competitive prices.

References

Advanced Research Projects Agency — Energy (n.d.). Biofuel production from kelp. https://arpa-e. energy.gov/technologies/projects/biofuel-production-kelp

Marine BioEnergy Inc. https://www.marinebiomass.com/

Office of Energy Efficiency & Renewable Energy. (n.d.). Biofuel Basics. https://www.energy.gov/eere/ bioenergy/biofuel-basics

Wikipedia. (n.d.). Biofuel. https://en.wikipedia.org/wiki/Biofuel

VI. Building, Food, and Other Sectors

Carbon Dioxide Capturing Processes

E arth's climate system is currently absorbing increasingly high levels of CO_2 into its atmosphere. The rising CO_2 concentration is destabilizing the climate, and the goal of achieving CO_2 neutrality by 2050 seems realistically unattainable. Even natural carbon sinks (billions of trees, the oceans, rocks) will not be able to remove CO_2 from the atmosphere fast enough and in the necessary quantities. For this reason, other attempts to solve the problem of removing CO_2 from the atmosphere by mechanical means are just as necessary as phasing out fossil fuels as quickly as possible.

CCS (Carbon Capture and Storage) and CCU (Carbon Capture and Utilization) are the two main technologies that can achieve this goal. While CCS involves the permanent storage of CO_2, CCU involves the conversion of captured CO_2 into valuable products such as fuels and other chemical products. These technologies do NOT eliminate the production of CO_2, but they will help reduce the CO_2 concentration in the atmosphere.

By capturing CO_2 during the combustion of fossil fuels and then storing it underground, up to 80% of the CO_2 produced could be permanently kept out of the atmosphere. Whether CCS technology can actually deliver on this promise is uncertain, and it remains the subject of various research and pilot projects. A decisive factor for feasibility is that CCS is very energy-intensive: CCS technology increases the consumption of raw fossil fuel materials by up to 40%. CCS can, therefore, only make a truly effective contribution to combating climate change if the stored CO_2 remains permanently and completely in its intended storage locations, i.e., in older

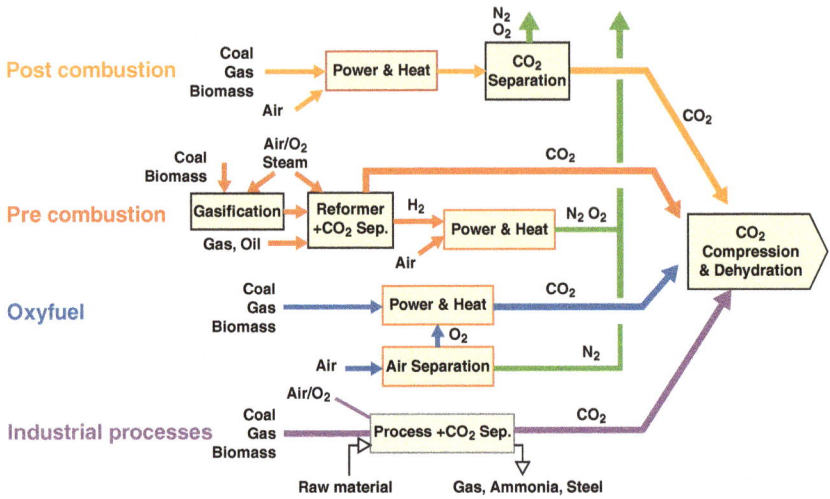

Figure 43.   Overview of CO2 capture processes and systems (IPCC, 2005).

gas or oil reservoirs that have already been exploited, in saline aquifers (deep-lying, saltwater-bearing rock layers), or in the oceanic subsurface.

The CCU process uses CO2 from industrial processes for further industrial purposes rather than releasing it into the atmosphere. The CO2 is used either in combination with power-to-gas plants to produce fuels or to produce basic materials for the chemical industry. CCU results only in a shift, not a reduction, in CO2 emissions. Also, the amount of energy (i.e., electricity) required to run this process is very high. Therefore, so long as this electrical energy does not come exclusively from renewable sources, the energy consumption for CCU leads to additional greenhouse gas (GHG) emissions.

References

Climate Council. (n.d.). What is carbon capture and storage? https://www.climatecouncil.org.au/resources/what-is-carbon-capture-and-storage/

Global CCS Institute. (n.d.). Understanding CCS. https://www.globalccsinstitute.com/about/what-is-ccs/

van der Meer, R., De Coninck, E., Helseth, J., Whiriskey, K., Perimenis, A., and Heberle, A. (2020). A method to calculate the positive effects of CCS and CCU on climate change. Zero Emissions Platform, July. https://zeroemissionsplatform.eu/wp-content/uploads/A-method-to-calculate-the-positive-effects-of-CCS-and-CCU-on-climate-change-July-2020.pdf

Wikipedia. (n.d.). Carbon capture and utilization. https://en.wikipedia.org/wiki/Carbon_capture_and_utilization

Direct Air Capture and Solar-to-Fuel

irect Air Capture (DAC) is an energy-intensive technology of the geoengineering industry. Geoengineering includes all technical procedures that interfere with the biochemical processes of the Earth. All the technologies, such as CCS and CCU, are criticized by fundamental climate change activists because they are mainly used by the fossil fuel industry to achieve targeted climate goals. However, from the point of view of a bridging or transitional technology that can prevent a climate catastrophe, geoengineering remains interesting and perhaps even necessary for the unavoidable transitional period. Today, DAC companies are still small (compared to their actual mission), but they are likely on the verge of explosive growth. A lot of money is flowing into DAC firms, with large sums coming from the fossil fuel industry.

In the DAC process, ambient air is drawn into a device, where it passes through a separator that binds the CO_2. This way of "purifying" air is certainly not new; it has long been used in submarines and space stations.

Solar-to-fuel refers to technologies that produce fuels from water, CO_2, and sunlight. Solar fuels can be produced in a carbon-neutral manner if the CO_2 required for the process is captured directly from the atmosphere. The cycle generated by this technology could soon be used to keep conventional aircraft running on carbon-neutral fuel. There are also plans to use solar fuels in other transport sectors, such as shipping and road traffic. In Switzerland, two companies are cooperating closely to put the cycle described (DAC and solar-to-fuel) into practice.

Scientists at ETH Zurich have built a plant that can produce carbon-neutral liquid fuels, known as "solar kerosene" (solar fuel), from sunlight and air. In this process, CO2 is taken directly from ambient air and split into carbon (C) and oxygen (O2) with the help of solar energy. This process produces synthesis gas, a mixture of hydrogen and carbon monoxide, which is then further processed into kerosene, methanol, or other hydrocarbons.

A research team led by Aldo Steinfeld, professor for renewable energy carriers at ETH Zurich, has been operating a miniature solar refinery on the roof of the ETH machine laboratory in Zurich for several years to test this process. The solar refinery is based on pure thermodynamics; it is climate-neutral because solar energy is used for production, and only as much CO2 is released during combustion as was previously removed from the air. Analyses of the entire process show that the fuel would cost 1.20 to 2 euros per liter if it were produced on an industrial scale. Desert regions with high solar resources would make particularly suitable production sites, and such plans are already being worked on around the world.

ETH Zurich did the groundbreaking scientific work, and now the spin-off company Synhelion is commercializing the technology to produce solar fuel from CO2. Given the high initial investment costs, solar fuel technology needs political support to ensure its ability to enter the market. Synthetic fuels (synfuels) are critical to opening up the world's largest markets. Aircraft, cars, trucks, and ships could all be converted to synthetic fuels that are chemically identical to their predecessors — without the need to retrofit vehicles, let alone decommission them.

Synhelion has recently announced that solar energy can be used to produce clinker cement, which contains about 40% clinker. Gianluca Ambrosetti, CEO and co-founder of Synhelion, said, "Our technology converts concentrated sunlight into the hottest existing solar process heat — beyond

1,500°C — on the market. We are proud to demonstrate together with CEMEX one specific industrially relevant application of our fully renewable, high-temperature solar heat."

The Swiss company Climeworks is one of the fastest-growing DAC companies in the world. Their pilot plant is located in an industrial area in Hinwil, near Zurich. Climeworks claims to be able to do the work of 36,000 trees with the footprint of a single tree. Many interested parties and prominent visitors (Forbes Magazine, Greta Thunberg) have already caught a promotional glimpse of the DAC machines, which look like oversized clothes dryers. Behind them are "balloons" that absorb the CO_2. In Hinwil, the CO_2 is transported to a beverage factory, where it adds the necessary carbonation to soft drinks; another portion is piped to a nearby greenhouse to improve fertilization. However, beverages and vegetables can only absorb a limited amount of CO_2. In 2017, the company opened another plant that takes advantage of Iceland's unique geology. The plant uses geothermal energy to capture carbon and then shoots the CO_2 into basalt rock for storage, where it turns to stone in less than two years. Climeworks has already partnered with Audi and Lufthansa; its biggest coup to date is a deal with a Norwegian consortium to build the first plant producing jet fuel without fossil fuels. If all goes according to plan, the planned plant will enable planes to fly on ether.

DAC has one major problem — economics. The relatively small amount of CO_2 in the atmosphere (0.04%) requires that enormous amounts of ambient air be pumped through the filters to capture significant amounts of CO_2. In turn, capturing the absorbed CO_2 from the filters requires huge amounts of thermal energy. Climeworks (and Global Thermostat, another DAC company) have managed to push the operating temperature below 100°C so that their plants can run on waste heat from industrial processes, of which there is no shortage. Otherwise, solar-powered pumps or collectors can provide the needed thermal energy. Ultimately, however,

DAC will consume significant amounts of energy to provide the required heat for the process. Under current conditions, it is not possible to economically operate the plants because it is still cheaper to pay carbon taxes than to capture carbon from the atmosphere.

Figure 44.   DAC facility by Climeworks. (Source: Julia Dunlop)

There have been many positive developments in DAC technology. A German consortium (DITF, Center for Solar Energy and Hydrogen, Institute for Energy and Environmental Research Heidelberg, Mercedes-Benz) believes that their process — called "Cora," which is based on a filter made of a special fabric — is energetically more favorable than the rather expensive process of the Swiss pioneer Climeworks. The latter process captures CO2 with a special sponge; when this is saturated, it must be replaced by a new sponge and regenerated in a heat chamber. DAC technologies costs are currently about 2–6 times higher than the desired cost of under $100 per ton of CO2. The theoretical advantages of the Cora process have yet to be verified in a pilot project.

References

CEMEX. (2022). CEMEX and Synhelion achieve breakthrough in cement production with solar energy. February 2. https://www.cemex.com/w/cemex-and-synhelion-achieve-breakthrough-in-cement-production-with-solar-energy

Climeworks. https://climeworks.com/

CO2-WIN. (n.d.). CORA — Separation of CO2 from air for power-to-X processes for sector coupling. https://co2-utilization.net/en/projects/chemical-and-biotechnological-reduction-of-co2/cora/

Furler, P. (n.d.). Synhelion. ETH Foundation. https://ethz-foundation.ch/pioneer-fellows/synhelion/

Rhode, E. (2021). Direct air capture pros and cons. Treehugger, April 26. https://www.treehugger.com/direct-air-capture-pros-and-cons-5119399

Synhelion. https://synhelion.com/

Wikipedia. (n.d.). Syngas. https://en.wikipedia.org/wiki/Syngas

Mechanical Trees

K laus Lackner, a German-American physicist, is director of the
Center for Negative Carbon Emissions (CNCE) at Arizona State
University. He has worked at the Stanford Linear Accelerator
Center, Los Alamos National Laboratory, and Columbia University's Earth
Institute. In 1995, Lackner proposed that a giant array of solar panels be
built by self-replicating robots in the desert areas of the White Sands
Missile Range. In 1999, Lackner was the first scientist to promote DAC.
He is sometimes referred to as the "father of DAC technology." However,
this is not quite true, as German-Swiss nuclear physicist Walter Seifritz
wrote in a letter to *Nature* as early as 1990: "If fossil fuels are to be used
on a massive scale and the greenhouse effect avoided, carbon dioxide in
the atmosphere must be reduced... [and/or disposed of] in empty natural
gas fields or the deep ocean." Seifritz also wanted to equip automobiles
with a CO_2 capture tank, but nothing came of that prophetic advice
either.

Lackner's latest invention in the field of DAC is the design of the so-called
"Mechanical Tree." DAC processes are usually based on the chemical or
physical separation of CO_2 from ambient air. Lackner's passive Direct Air
Capture (PDACTM) technology avoids energy-intensive processes, making
the technology relatively low-cost, scalable, and commercially viable by
using wind to power the system. Mechanical Tree technology is being
promoted as a key solution for companies and legislators pursuing carbon
reduction strategies and aiming for carbon neutrality through technological
CO_2 removal. A Mechanical Tree is, according to Lackner's company, "up
to a thousand times more efficient at removing CO_2 from the air as a
natural tree."

The "tree" itself is a column about 10 meters high containing "sorbent tiles" that extend and retract in a continuous collection and regeneration cycle. Lackner's company, Carbon Collect, claims that energy is only needed for processing by sequestering the captured CO_2. Lackner himself has pointed out the real purpose of his invention: "Oil companies' entire business model will fall apart if carbon dioxide cannot be recovered from the atmosphere and environmental carbon constraints become more severe." In order to maintain the oil industry's business model, increasingly stringent environmental regulations must be met; this can be done with DAC processes such as "Mechanical Trees."

Basically, in energy policy terms, DAC is a truly welcome way for all investors and politicians who want to use carbon engineering technology to enable the fossil fuel industry to continue, intentionally or not.

David Keith is the founder of Carbon Engineering, a Canadian company. Keith believes that DAC has lower adaptation costs than conventional mitigation options, and it is, therefore, "optimal to pollute more when it is possible to clean up afterward than when it is not." Sentiments like these have already appealed to investors with fossil fuel backgrounds, such as BHP, a British-Australian mining and petroleum conglomerate, or Chevron, an oil company, or Occidental, which wants to be the first oil company to sell "carbon-negative oil." Commenting on the move, Fiona Wild, vice president of BHP, said, "This is about recognizing that climate change poses significant risks to all economic sectors. Climate change is no longer seen as a fringe issue. It's a business risk that requires a business response." If Occidental buys CO_2 from Carbon Engineering and squeezes it under the oil, it can claim so-called "45Q tax credits" — increased US tax credits for CO_2 use and storage, which are expected to help power plants and industrial plants decarbonize.

Together with the investment firm Rusheen Capital Management, Carbon Engineering formed a joint venture called "1pointfive" in August 2020.

"1pointfive" plans to build the world's largest DAC plant. Located in Texas, the plant is expected to be capable of capturing one million metric tons of CO_2 annually once it comes online in late 2024. Aircraft manufacturer Airbus has already secured 400,000 tons of CO_2 "credits" for compensating the growing carbon footprint of their still-growing fleet.

Graciela Chichilnisky and Peter Eisenberger are the founders and CEOs of Global Thermostat, another major DAC player. Both worked with Lackner at Columbia University's Earth Institute. In an interview with Forbes KPMG, Chichilnisky said: "The idea here is to protect our earth from global warming and by absorbing the carbon emissions, we can reduce the damage to our environment by reversing the process." Back in June 2019, ExxonMobil signed a contract with Global Thermostat to build an adequate number of DAC plants. Until then, the rather small start-up had built only a handful of pilot plants with a capacity of 4,000 tons of CO_2 per year. When ExxonMobil signed the contract, Chichilnisky and Eisenberger foresaw a gigantic development push for their technology. This partnership ended in 2023, and ExxonMobil started working on an "in-house" DAC demonstration project. The global oil industry is always finding ways and means to continue production and, thus, counteract the fight against climate change in its own way.

References

Arizona State University. (2019). Powerful „mechanical trees" can remove CO_2 from air to combat global warming at scale. *ASU News*, April 29. https://news.asu.edu/20190429-solutions-lackner-carbon-capture-technology-moves-commercialization

Carbon Collect. https://carboncollect.com/

1PointFive. (n.d.). 1PointFive holds groundbreaking for world's largest direct air capture (DAC) plant. https://www.1pointfive.com/1pointfive-holds-groundbreaking

123HelpMe. (n.d.). Global thermostat case study. https://www.123helpme.com/essay/Global-Thermostat-Case-Study-723516

Carbon Engineering. https://carbonengineering.com/

EurekAlert! (2022). Carbon Collect unveils MechanicalTree™ in partnership with Arizona State University. April 22. https://www.eurekalert.org/news-releases/950580

Habib, M.A. (2022). Undoing the damage. *Dhaka Tribune*, November 29. https://www.dhakatribune.com/epaper/299302/undoing-the-damage

Keith, D. (2001). Geoengineering and carbon management: Is there a meaningful distinction? https://gis.huri.harvard.edu/files/tkg/files/41.keith_.2001.geogineeeringandcarbonmanagment.f.pdf

Keith, D., Minh, H.-D., Stolarof, J.K. (2006). Climate strategy with CO2 capture from the air. *Climatic Change*, **74**(1–3): 17–45. https://shs.hal.science/file/index/docid/30417/filename/Keithetal-AirCapture.pdf

Malm, A. and Carton, W. (2021). Seize the means of carbon removal: The political economy of direct air capture. BRILL, March 16. https://brill.com/view/journals/hima/29/1/article-p3_1.xml?language=en

National Energy Technology Laboratory. (2021). DOE invests funding to decarbonize the natural gas power and industrial sectors using CCS. *Carbon Capture*, November. https://netl.doe.gov/sites/default/files/publication/NETL-November-2021-Carbon-Capture-Newsletter.pdf

The Royal Society. (2009). Geoengineering the climate: Science, governance and uncertainty. RS Policy document 10/09, September. https://royalsociety.org/~/media/royal_society_content/policy/publications/2009/8693.pdf

US Department of Energy. (1999). Carbon sequestration research and development. Office of Scientific and Technical Information, December 31. https://www.osti.gov/servlets/purl/810722-9s7bTP/native/

Wikipedia. (n.d.). Klaus Lackner. https://de.wikipedia.org/wiki/Klaus_Lackner

What Is the Cost of Direct Air Capture?

A great deal of energy must be allocated to DAC technology if it is to have a real impact on climate. The math is simple: To keep global warming below 2°C, 30 gigatons of CO_2 per year must be removed from the atmosphere. To achieve this goal, DAC would presently consume more than half of the electricity produced globally today. The total cost per ton of captured CO_2 is also daunting. Climeworks has reached a price of $600 per ton in Iceland (including sequestration), and Carbon Engineering has quoted their product at $230, while Global Thermostat boasted a price of $100, soon to be cut in half.

According to Christian Parenti, an American investigative journalist, removing 40 gigatons of CO_2 using known DAC technologies would require a staggering sum of $12 trillion — annually! He pointed out, correctly, that other scenarios could be even more costly: rapid sea level rise leading to flooded cities (New York City, Mumbai), devastating wildfires, and the whole gamut of potential consequences of global, economic, and social collapse. In this equation, removing atmospheric CO_2 by all available means is a preferable option to prevent catastrophe.

To achieve these goals, 10 to 20 million DAC machines must be installed each year. This may be within the realm of feasibility, given that more than 70 million cars are already produced each year. If we assume that governments will invest 2% of global GDP in DAC machines each year after 2025 — in the manner of a new "Manhattan Project" — we could probably remove 570 to 840 Gt of CO_2 from the atmosphere by 2100. This represents a total of roughly 10 to 15 times our current annual emissions.

The result would be a temperature reduction of about 0.1 to 0.2°C by 2100, but that is a fraction of the planet's expected warming scenario.

DAC companies are currently being embraced by carbon-producing firms (oil companies) and capital investment firms because DAC has a clear advantage over existing CO2-avoiding alternatives such as solar, wind, geothermal, and hydropower. Theoretically, it offers the chance of near-total control over the terms of carbon disposal. If all goes according to the plan of the old economy, "green" alternatives will not displace fossil fuels but merely supplement them. Considering that, something fundamental may have to change. Under a more aggressive and essential approach, policymakers could intervene quickly and set ambitious targets to be met by carbon producers. Touting natural gas and nuclear power plants as green solutions is a helpless reaction to rising CO2 emissions. DAC could play an important role in bringing CO2 numbers down, but only with an "all-out" effort by building a huge number of large plants. According to an IEA (International Energy Agency) study, building 32 large-scale DAC plants per year until 2050 will be necessary to meet Net Zero goals. This translates into investments totaling several trillion USD, according to the IEA DAC report of 2022.

References

International Energy Agency. (2022). Direct Air Capture 2022. Report. https://www.iea.org/reports/direct-air-capture-2022

Keating, C. (2023). What will scale direct air capture? A 75 percent price drop, report says. Green Biz Group, June 8. https://www.greenbiz.com/article/what-will-scale-direct-air-capture-75-percent-price-drop-report-says

Lebling, K., Leslie-Bole, H., Byrum, Z., and Bridgwater, L. (2022). 6 Things to know about direct air capture. World Resources Institute, May 2. https://www.wri.org/insights/direct-air-capture-resource-considerations-and-costs-carbon-removal

Malm, A. and Carton, W. (2021). Seize the means of carbon removal: The political economy of direct air capture. *Historical Materialism*, **29**(1), 3–48. https://lucris.lub.lu.se/ws/portalfiles/portal/96341244/HM_DAC.pdf

Williams, E. (2022). The economics of direct air carbon capture and storage. Global CCS Institute. https://www.globalccsinstitute.com/wp-content/uploads/2022/07/Economics-of-DAC_FINAL.pdf

Sustainable Food

The World Population 2020 fact sheet published by the US Population Reference Bureau shows that the world's population is likely to increase from 7.8 billion in 2020 to 9.9 billion in 2050, meaning that about 2 billion more people will need to be fed. Two-thirds of these people are projected to live in cities by that point. As incomes rise in the Global South, people will increasingly want to consume more resource-intensive and animal-based foods. There is a huge gap between the amount of food we produce today and the amount that will be needed in 2050. The food industry is literally begging for innovations or substitutes, especially for animal foods.

World Population Indicators: 1950, 2020 and 2050	1950	2020	2050
Total (billions)	2.5	7.8	9.7
Annual growth rate	1.8%	1%	0.5%
Annual increase (millions)	77	81	43
Sex ratio (males/100 females)	100	102	101
Percent in Urban Areas	30%	56%	68%
Number of megacities	1	33	48
Life expectancy at birth (years)	47	73	78
Infant mortality (deaths/1,000 births)	140	29	15
Total fertility rate (births per woman)	5	2.4	2.2
Percent under 15 years	34%	25%	21%
Percent 65 years and older	5%	9%	16%
Potential support ratio (15-64 per 65+)	12	7	4
Total immigrants (millions)	72*	275	400**
Total refugees (millions)	3*	26	40**

Figure 45. World population indicators. (Source: UN Population Division, UNHCR/UNRWA)

Agriculture is one of the biggest contributors to global warming, primarily through methane released from livestock and rice farms, nitrous oxide from fertilized fields, and CO2 from deforestation (and conversion of other land) to grow crops or raise livestock. In addition, agriculture consumes much of our water supply; fertilizers and manure destroy fragile ecosystems and are one of the many causes of global biodiversity loss.

Today, our food system emits about 9 gigatons of CO2 each year. The goal of achieving Net Zero by 2050 requires a reduction in agricultural emissions to a total of 2 gigatons. Our current system is based on industrial agriculture, industrial fertilization, rice farming, cattle, lamb, goat, and pig fattening. Nitrogen fertilization causes nitrous oxide to be released into the atmosphere, where it traps three hundred times more heat than CO2. To meet the goals of the 2015 Paris Agreement, all agricultural production must become sustainable. We must improve soil health, increase soil carbon content, and reduce food waste — about one-third of all vegetables and fruits grown are thrown away. At the same time, it is crucial to stop the conversion of existing forests to agricultural land.

For the food industry, a sustainable future means that the type, composition, and quantity of food people consume will have to change significantly. To achieve this, two very different gaps must first be closed: (a) a more than 50% food gap between the amount of calories produced in 2010 and the expected calorie demand in 2050, and (b) a land gap of more than 500 million acres between the amount of agricultural land in 2010 and the volume of land that will be required by 2050.

How can this life-threatening situation be resolved? One way is to increase yields on less productive soils by using high-tech farming systems. Organic farming (with cover crops, mulch, and compost) can significantly reduce the use of water and chemicals, allowing us to increase yields many times over. We also need to reduce our collective overconsumption of calories and animal protein, especially beef. Beef produces 20 times more GHGs than vegetable proteins (beans, peas, lentils) throughout its production chain. Limiting meat consumption to 1.5 burgers per person per week would cut the GHG reduction gap in half.

Ruminants are responsible for half of all agricultural GHG emissions, with methane from cows accounting for the largest share. Trials of chemicals and other natural feed additives to inhibit gas emissions have led to promising ~30% reductions. Improved manure management, reduced emissions from fertilizers, and better rice management, as well as increased energy efficiency in agriculture by switching to non-fossil fuels, would contribute significantly to the radically lower agricultural carbon footprint needed for Net Zero by 2050 targets.

Global food waste is estimated at $1 trillion per year. Measures to mitigate this waste include better monitoring of food chains, improved refrigeration and storage, and better expiration date labeling. Other necessary actions include better marketing of plant-based foods, better meat substitutes, and the implementation of policies that encourage widespread consumption of plant-based foods. This could be done, for example, by highlighting the carbon footprint of food choices on menus or cash register receipts.

Adapting to climate change requires the cultivation and specialized breeding of crops that can cope with higher temperatures, changing production systems, and the introduction of better irrigation systems. Otherwise, global crop yields will most likely decline dramatically due to climate change.

Since most marine stocks are already overfished, all subsidies that led to this process must be consistently phased out. At the same time, productivity, feed quality, disease control, and environmental sustainability in aquacultures must be improved.

Consumers are increasingly aware of the ecological and ethical benefits of a vegetarian or vegan diet. Nevertheless, consumers still prefer to continue eating animal products out of habit, taste preferences, convenience, and price ("industrial meat is cheap!"). However, if you take a closer look at the shelves in grocery stores, you will find more and more products that are rich in protein and derived from animal-free raw materials. Beyond Meat, a US company, makes burgers composed of peas and brown rice, among other ingredients, and has already signed a strategic agreement with McDonald's. Their goal is to achieve price parity with beef by 2024. Upside Foods (formerly Memphis Foods) makes lab-grown meatballs from animal stem cells (obtained through a painless biopsy). The company also makes lab-grown poultry, which may explain why its investors include industry giants like Tyson Foods and Whole Foods.

Protein is an essential building block of human nutrition. It is crucial for building, maintaining, and repairing bones, hair, and other tissues. For this reason, animal-free foods often focus on the production of protein-rich components. The goal is to take the animal out of the equation but still produce functional animal proteins. Several companies around the world are working on related product portfolios to produce dairy or animal proteins from plants or mushrooms; for example, Emergy Foods of Boulder, Colorado, makes mushroom-based steaks, which are not made directly from mushrooms but from the mycelium — the fast-growing rhizome or fungal tissue that precedes mushroom development.

Mushrooms are ideal food due to their high protein content (20–30%). Mushroom-based foods have a long history, at least in Asia (koji, soy sauce). For example, the start-up company Smallhold specializes in growing, packaging, and selling mushrooms. The consumption of insects is already widespread in Asia, Mexico, and some African countries. They are rich in fat, protein, vitamins, fiber, and minerals and can be produced many times more sustainably than farm animals. Although many people are disgusted by the idea of eating spiders and flies, it is possible to present them in a much more attractive and tasty context. One of the world's largest producers in this field is the French company Ynsect. It operates the largest vertical farm in the world and has recently received $400 million in investment money to help feed the world with — insects. Vertical farming is a new technology where crops and other foods are grown in vertically stacked layers.

Algae, fish, and seafood imitations are the next big thing in food production. Algae (e.g., kelp) contain high levels of omega-3 fatty acids, an important nutrient. Kelp is the fastest-growing "protist" — a very large brown algae — on the planet. It has no roots and is easy to grow. Kelp harvests energy from the sun through photosynthesis and does not feed on other organisms. Kelp tastes good and is a perfect way to provide delicious seafood to vegetarians and vegans.

Figure 46. Kelp Forest tank, Monterey Bay Aquarium, Monterey County, California, US. (Public Domain)

Dairy products that are traditionally derived from cow's milk can also be made from plants. They taste just as delicious as widely available cow's milk products. Although it is a "biotechnological challenge" to mimic the proteins of cow's milk to produce lab-grown dairy products, some companies claim they have already succeeded. "Perfect Day," a Silicon Valley-based startup, is already bringing its high-priced products to market.

SNACT, a UK-based eco-startup, is helping to prevent food waste. They sell a snack product line where each bar saves a banana from disposal. In addition, each bar is wrapped in an innovative, compostable film made by TIPA. TIPA is based in Israel and has operations in Europe and the US. Their packaging materials are fully biodegradable but meet the same requirements as everyday conventional plastic.

References

Astanor. (n.d.). Cases. https://astanor.com/cases/

Beyond Meat. https://www.beyondmeat.com/

Considerate Consumer. (n.d.). Animal friendly food options. https://www.considerate-consumer.com/animal-considerate-food

Meati. https://meati.com/

Moore, D. and Chiu, S. (2001). Fungal products as food. In *Bio-Exploitation of Filamentous Fungi*, S. B. Pointing, S.B. and Hyde, K.D. (eds.). Fungal Diversity Press, Hong Kong, pp. 223–251.

Motif Foodworks. https://madewithmotif.com/

New Wave Foods. https://www.newwavefoods.com/

Perfect Day. https://perfectday.com/

Population Research Bureau. (n.d.). 2020 world population data sheet. https://interactives.prb.org/2020-wpds/

Protix. https://protix.eu/

Seafood Nutrition Partnership. https://www.seafoodnutrition.org/

Smallhold. https://smallhold.com/

TIPA. https://tipa-corp.com/

Upside Foods. https://upsidefoods.com/

Webb, S. (2022). Customers shun red meat after carbon footprint added to weekly shop receipts. *The Independent*, January 18. https://www.independent.co.uk/climate-change/news/oda-red-meat-carbon-footprint-b1988091.html

Ynsect. https://www.ynsect.com//

How to Build Sustainably?

C ities cover about 3% of the Earth's surface, but half the world's population now lives in them. Over the next 30 years, an estimated one million people per week will move into the world's cities. In Asia alone, an additional 1.1 billion people will move into cities over the next 20 years. According to the United Nations (UN), 68% of the world's population will live in cities by 2050.

Cities, especially those located in coastal areas, are among the places most likely to feel the acute effects of climate change. That is one reason why, in 2017, the mayors of 25 cities around the world committed to becoming climate-neutral by 2050. By 2022, this network, called "C40," already includes close to 100 cities, including Austin, Accra, Barcelona, Boston, Buenos Aires, Cape Town, Caracas, Copenhagen, Durban, London, Los Angeles, Melbourne, Mexico City, Milan, New York City, Oslo, Paris, Philadelphia, Portland, Quito, Rio de Janeiro, Salvador, Santiago, Stockholm, and Vancouver. Anne Hidalgo, Mayor of Paris and C40 Chair, said, "Mayors of the world's great cities are shaping the century ahead and paving the way for a better, healthier, and greener future. Mayors do what they must do, not what they can, and these plans and policies are an excellent example of our state of mind."

A lot of energy is needed to construct, maintain, and operate buildings, while the heating, cooling, and operation of buildings produce many climate-damaging gasses. How can the dependence of buildings on fossil fuels be reduced? The answer is quite simple: build passive houses with better thermal performance because the main source of CO_2 emissions is heating. A well-known example of this type of building is the Energon office building in Ulm, Germany, designed by Stefan Oehler and measuring approximately 5,700 square meters. Built in 2002, it is one of the first of more than 30,000 buildings that are passive houses today.

In the architecture of the groundbreaking Energon building, energetic and ecological principles were consistently taken into account in the design of the structure and the façade, and five principles were followed for the construction and operation of the building:

- exceptionally good thermal insulation of the entire building shell,
- extensive tightness of the building shell,
- controlled ventilation with heat recovery,
- extensive use of passive solar and internal heat gains while minimizing internal loads, and
- meeting the necessary energy requirements with a high level of efficiency, using regenerative energies to the greatest possible extent.

Key features of the Energon building include thermal component activation for heating and cooling as well as regenerative cooling by means of geothermal probes. Forty of these geothermal probes, each 100 meters deep, have water flowing through them as a heat transfer medium. They give off heat to the ground in summer and absorb heat from the constant earth temperature of 10°C in winter. A photovoltaic system on the roof of the building and another system on the neighboring parking garage handle around 70% of the building's primary energy requirements.

Underground ducts around the building draw in air and heat it during the winter. In summer, they cool the system using the aforementioned probes, penetrating to a depth where the Earth's natural temperature can be used to cool the air above. The building uses 75% less energy for heating and cooling than a normal office building. In addition, the construction method employed is no more expensive than that of a conventional air-conditioned office building.

Even before they are put to their intended use, buildings have an enormous carbon footprint. Emissions occur in the construction process and are already contained in the building materials. Currently, steel and concrete production account for about 16% of global CO2 emissions. Using wood instead of steel and concrete could prevent the release of 20–30% GHG emissions, according to Finnish research conducted by Aalto University. Trees absorb CO2 from the atmosphere, and their use as lumber could then sequester that CO2 for decades. UPM, for example, a Finnish "forest industry pioneer," focuses its portfolio on renewable raw materials such as pulp, wood, and a whole mix of innovative, sustainable materials for construction.

The Lee Kuan Yew Centre for Innovative Cities (LKYCIC) — a research institute at the Singapore University of Technology and Design — focuses on the study of cities, urbanization issues, and the use of new technologies and designs. Professor Cheong Koon Hean — current chair of the institute — was featured in the BBC program "2045: Memories of the Future." She played an active role in the urban planning and development of Singapore, one of the "smart cities" of our future. A "smart city" encompasses new ideas and comprehensive concepts for urban spaces. Such cities are made in more efficient ways, making them more climate-friendly and livable through the use of modern technologies. To achieve these goals, new types of products, services, and infrastructural elements are needed, supported by highly integrated and networked information and communication technologies. The central concern of a smart city is to increase the efficiency, effectiveness, resilience, and sustainability of the overall system.

A city is considered "smart" when investments in human and social capital, as well as traditional (transportation) and modern (communications) infrastructure, promote sustainable economic growth and a high quality of life, while natural resources are managed wisely through participatory governance.

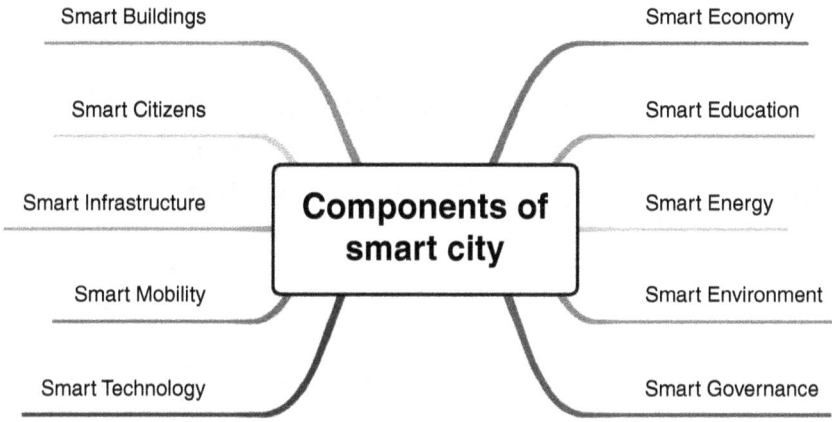

Figure 47. Components of a smart city. (Source: JudyMiao, CC BY-SA 4.0, https://commons.wikimedia.org/w/index.php?curid=126466410)

A Smart Sustainable City (SSC) is a city that dramatically improves its social, economic, and environmental (sustainability) outcomes and responds to challenges such as climate change, rapid population growth, and political and economic instability by:

- meeting the needs of present and future generations regarding economic, social, and environmental aspects,
- using cooperative leadership methods to ensure accountability and responsibility,
- working interdisciplinarily and across urban systems, and
- improving the use of data and modern technologies to improve quality of life and efficiency of urban operations and services.

By implementing these goals, SSCs can provide better services and a higher quality of life to citizens of the city and those who interact with it, now and in the foreseeable future.

Making cities smarter, more efficient, and more sustainable for their residents has never been more important. Key areas include improved basic technologies, buildings and construction, improved energy procurement, management, and use, as well as smart water and waste management. Smart, sustainable, and resilient cities bring a wealth of benefits to their residents and investors. While Asia remains ahead of the curve, aging cities, particularly in the West, are under pressure to rapidly modernize their outdated infrastructure.

What is a resilient city? Urban resilience is the ability of a large group of people or communities within a city to survive, adapt, and thrive, regardless of the types of chronic stresses and/or acute shocks or disasters they experience — according to Michael Berkowitz, founder of Resilient Cities Catalyst, a global nonprofit organization that helps cities and their partners overcome the most difficult challenges and create thriving and economically resilient communities.

Paris Mayor Anne Hidalgo worries about flooding, heat, and air pollution. In Paris, it all started with an approach to reduce summer heat in the city after record temperatures of 42°C were recorded in July 2019. New measures include rainwater capture and storage, green roofs, additional planting, shade coming from new shrubs and trees ("urban forests"), and better access to green spaces. These measures have very quickly led to lower temperatures and a reduced risk of flooding. Mayor Hidalgo is also aiming for more than 600 kilometers of bike lanes and wants to make downtown Paris car-free, which has led to angry reactions from cab drivers, restaurant owners, and delivery drivers. The resolute mayor has also had parts of the expressway on the right bank of the Seine shut down and turned into a park. As part of her campaign, she plans to open up the Seine, which is once again safe from a health point of view, for swimming competitions during the 2024 Olympics and to the public afterward.

England's capital, London, the largest metropolitan area in the United Kingdom, has taken a similar approach, but with some notable differences. The city's Repowering London campaign aims to provide London with sustainable, community-managed energy because "putting people at the heart of the energy system is the key to fighting the climate emergency." Legal entities have been established to give people control over their energy choices and to plan an active role in the energy transition. Shares can be purchased by residents, businesses, and municipalities. Each member gets a vote, regardless of the size of their investment, to ensure a truly democratic community. Many projects under this umbrella aim to create smart local energy systems. In many cases, small, discarded parcels of land are used to create so-called energy gardens with solar panels. In this context, the development and use of previously unused land and the joint action further contribute to the community's sense of belonging.

References

Alexandre G. and Rizhlaine F. (2020). Urban forests in Paris: Two projects gave up by Anne Hidalgo, three new sites presented. Sortiraparis, November 9. https://www.sortiraparis.com/news/in-paris/articles/225782-urban-forests-in-paris-two-projects-gave-up-by-anne-hidalgo-three-new-sites-presented/lang/en

C40. (2016). Cities see economic potential in partnering with businesses on climate action. October 4. https://www.c40.org/tag/smart-cities/

C40. (2017). 25 cities commit to become emissions neutral by 2050 to deliver on their share of the Paris Agreement. Press release, November 12. https://www.c40.org/news/25-cities-emissions-neutral-by-2050/

International Telecommunication Union. (n.d.). Digital transformation for people-centred cities. https://www.itu.int/en/ITU-T/ssc/Pages/default.aspx

Lee Kuan Yew Centre for Innovative Cities. https://lkycic.sutd.edu.sg/

Repowering London. (n.d.). Trailblazers in community energy. https://www.repowering.org.uk/our-story/

Resilient Cities Catalyst. https://www.rcc.city/

Sustainable Smart Cities and Territories International Conference. https://ssctic.net/

United Nations Economic Commission for Europe. (n.d.). Smart sustainable cities. https://unece.org/housing/smart-sustainable-cities

United Nations Human Settlements Programme. (2022). World cities report 2022: Envisaging the future of cities. https://unhabitat.org/wcr/

YouTube. (2021). Smart & Sustainable Cities (SSC) Forum promo. November 24. https://www.youtube.com/watch?v=7sAYfCoy64E

Preventive Measures

Most climate scientists agree that by 2050, perhaps even sooner, about 90% of the world's megacities will be at risk from rising sea levels and consequent flooding. Homes and other critical infrastructural elements could be affected, and millions of lives will be threatened. The solution to this problem could lie in harnessing the ocean areas offshore from cities. Hong Kong-based Oceanix, run by Collins Chen in New York, plans to extend coastal cities out to sea as so-called "floating cities." The idea is to build houses and infrastructure on floating platforms at sea.

The floating platforms are being designed by the Center for Ocean Engineering at the Massachusetts Institute of Technology (MIT), and the complex above-water structures by a group of architects from New York City and Copenhagen. The buildings will be constructed of wood from sustainable forests, along with greenhouses, vertical farms, and underwater gardens. Energy will be supplied from renewable sources such as wind and solar power, coupled with wastewater and waste recycling facilities, as well as desalination plants for drinking water. This concept is intended to lay the foundation for "sustainable cities for the future." The modular concept will be able to change and adapt over time, evolving from a small village of perhaps 2–300 residents to a city of 10,000. A prototype is scheduled for completion in 2025, presumably off the coast of Busan, a South Korean city of 3.4 million people.

At a time when the weather has become dangerously unpredictable, with the risk of devastating floods increasing worldwide, simple and easy-to-implement solutions are needed. To combat flooding, especially in cities, Beijing landscape architect Kongjian Yu proposes so-called "sponge cities." The concept of sponge cities is to create citywide systems of ponds, wetlands, and parks that retain rainwater. Yu, founder of Turenscape, one of China's largest landscape architecture firms, explains, "Since ancient times, Chinese cities along the Yellow River with monsoon climates have used ponds to manage flooding and stormwater. So we know these approaches worked for over 2,000 years because these cities survived."

The key components of Yu's Sponge City concept use green infrastructure to capture rainwater at the source where the water falls. Yu sees great opportunities for landscape architects in the near future: "We have an opportunity to build up our approach. Landscape architects can solve these problems — not with concrete pipes and cisterns — but with nature." Yu's opinion on New York City's plans to build a huge concrete wall and large cisterns to store water is clear: "Cisterns are unsustainable."

The City of New York plans to build a series of landscaped concrete levees around the southern tip of Manhattan, known as the "Big U." The existing 60-acre East River Park will be buried under a man-made fill, and its new, higher edge will become a wall holding back the East River, which, like the Atlantic Ocean, is projected to rise about 75 centimeters by 2050. The park will provide flood protection up to 5 meters above the current sea level, protecting 100,000 residents along the East Side and Franklin Delano Roosevelt (FDR) Drive. The new, elevated park will cost close to $1.5 billion, which has upset more than a few residents. Beyond that, other key questions remain: "What will happen if sea levels rise and continue to rise? What will happen after 2050?"

References

Green, J. (2021). Kongjian Yu defends his sponge city campaign. *The Dirt*, April 8. https://dirt.asla.org/2021/08/04/kongjian-yu-defends-his-sponge-city-campaign/

Liew, M.E. (2022). Inside the world's first climate resilient floating city in South Korea. GovInsider, May 5. https://govinsider.asia/intl-en/article/inside-the-worlds-first-climate-resilient-floating-city-in-south-korea-oceanix-busan

Oceanix. https://oceanixcity.com/

Rebuild by Design. (n.d.). Project pages: The big U. https://rebuildbydesign.org/work/funded-projects/the-big-u/

Rethinking the Future. (n.d.). 5 Ideas that will shape the Future of Cities. https://www.re-thinkingthefuture.com/narratives/a4891-5-ideas-that-will-shape-the-future-of-cities/

Turenscape. https://www.turenscape.com/en/home/index.html

Vivid Futures. (n.d.). Articles. https://vividfutures.co.uk/news

Urban Gardening

C ities are also turning to other nature-based solutions to store carbon in an effort to counteract rising temperatures. Medellin in Colombia, for example, has created 30 green corridors along streets and waterways with thousands of trees and shrubs. This has reduced the local average temperature by more than 2°C. Medellin, as well as other densely populated cities in hotter climates, suffers from the urban heat island effect, in which materials such as concrete absorb solar energy and redirect it back into cities, says Benz Kotzen, associate professor in the School of Design at the University of Greenwich. One way to counteract this effect is to introduce nature-based solutions such as green roofs and "living walls." "Green infrastructure solves a lot of problems. It increases biodiversity, has a cooling effect from the evaporation of vegetation, and it can absorb some particulate pollution in the air," says Kotzen.

Urban gardening is another way to keep the local climate fresh and healthy. This concept can be divided into two segments: container gardening and rooftop gardening. The former is suitable for people with small terraces, yards, or balconies, while the latter is traditionally used for growing vegetables and larger plants. Urban gardening has been practiced since time immemorial, but especially during wartime, e.g., the "Victory Gardens" during World War I and World War II in the US. After 1945, half of the bombed-out German city centers were used for growing fruits and vegetables.

Figure 48. "Victory Garden," US War Food Administration, Hubert Morely, 1945. (Public Domain)

All you need to explore your gardening skills is enough sunlight (6–8 hours a day is ideal), good soil, and water. The Food and Agriculture Organization (FAO) of the United Nations defines urban gardening this way: "An industry that produces, processes, and markets food and fuel, largely in response to the daily demand of consumers within a town, city, or metropolis, on land and water dispersed throughout the urban and peri-urban area, applying intensive production methods, using and reusing natural resources and urban wastes to yield a diversity of crops and livestock."

Think globally — act locally. This slogan from an almost forgotten era still holds true.

References

Brown, J. (2021). How cities are going carbon neutral. BBC Future, November 16. https://www.bbc.com/future/article/20211115-how-cities-are-going-carbon-neutral

International Association for the Study of the Commons. (n.d.). Urban gardening. https://iasc-commons.org/cs-urban-gardening/

Wikipedia. (n.d.). Think globally, act locally. https://en.wikipedia.org/wiki/Think_globally,_act_locally

VII. How Change Will Succeed

We Must Change!

O ne hundred years ago, Nikola Tesla had the idea of making energy from the cosmos available to the entire planet. His ideas of a cosmic energy supply were rather dreamy, but there is one energy source that supplies the Earth with energy constantly and free of charge — our Sun. The sun releases so much energy that after an approximately 8-minute, light-speed flight through space, the solar energy that hits the Earth in one hour would be sufficient to fully cover our energy consumption of one year. Obviously, it is not possible to harness all of the sun's energy, but if we could shift our energy production to solar, wind, hydro, and thermal, we would be on a much better path to weathering the coming climate crisis. All we have to do is change, but it is very difficult to make real changes in our own lives. Those who have already gone through such changes know that, and those who have not will have to experience it. It takes will power and a lot of energy to change. The road will be long and full of obstacles. We will all be forced to go through the necessary transition; in fact, we are already in it. This is the new age of enlightenment, and we are here to fight for it, together.

- We are living in a life-threatening, critical time for our species.
- We need a colossal collective effort to meet the demands of climate change.
- Do not think we have much time for change. We do not.
- The coming decade is ultimately the last chance for change.
- Either you are in, or we are all out.

Electric vehicles (cars, buses, etc.) and hydrogen-powered aircraft are good examples of a new green world. However, these devices will not save the planet from the imminent climate crisis. Much more needs to be done on a personal level; we all need to make a lot of changes. To achieve Net Zero status by 2050, we need to significantly reduce our high-carbon-dioxide (CO_2) activities. Big gas-guzzling SUVs, daily servings of red meat (burgers), and cheap long-haul flights must all be abandoned for good. However, the coming change will demand much more from each and every one of us, especially for those living in the Global North. We all feel it. This is the elephant in the room that no one wants to acknowledge: We all need to change as soon as possible.

We will never again live as we did before the COVID-19 pandemic or the Ukraine war. A greener society needs greener people. We need more awareness, including mindfulness; we need to recognize our sense of responsibility. Focused thoughtfulness is a great challenge to our Paleolithic minds. We are facing the opportunity for a new age of enlightenment ("Enlightenment 2.0"). These are troubled times, and the transition to a sustainable and resilient society is the defining challenge of our time.

What are the alternatives? Laziness? Boredom? Another war? Civil wars? Or will we all be able to pull together and turn our heads and minds toward other goals? Lorraine Whitmarsh, MBE, a British Professor of Environmental Psychology and head of the Centre for Climate Change and Social Transformation (CAST), states "people underestimate just how much behavior change is going to be required." In a February 2022 press release, Professor Whitmarsh announced funding for the five-year Advancing Capacity for Climate and Environment Social Science (ACCESS) program, saying, "We know that radical social and behavioral change is needed to tackle climate change and other environmental issues, and this funding will champion and apply social science insights to achieve this change. ACCESS will bring together researchers, policymakers, and the broader society to find fair and effective solutions to environmental challenges while improving well-being and prosperity."

Professor Whitmarsh's holistic approach to inquiry begins with asking questions:

- From which areas of life do the climate gasses caused by individual humans originate in the first place (transport, food, housing, etc.)?
- To what extent is climate change affected by our purchasing behavior (cars, clothes, electronics, etc.)?
- To what extent do food production and land use need to be reconsidered, especially with regard to the possible relocation of food production abroad?
- What are the policy options (comprehensive information, taxes, laws) to change people's behavior?
- How must/can politics, business, and people act together for this common purpose?

In her practical studies, Professor Whitmarsh looks for ways to encourage people to adopt more sustainable, climate-friendly behavior. In tests, she has been able to demonstrate that, for example, lower electricity consumption by individual households can be achieved relatively easily by offering "incentives" in the form of lower tariffs for green electricity, for example, to a specific population group. Once this tariff is accepted by about 20% of the population (and who would not want to pay less for their electricity?), the rest of the population will automatically follow suit. This tactic of "persuasion" can probably be applied in any area, but you have to find the right "catalyst" to convince people in each case.

As director of the CAST program, Professor Whitmarsh collaborates with other scholars to develop the Principles of Practical Change. The four principles of the CAST program are:

- DEVELOP and expand the VISION of a future low-carbon society and ways to achieve the goals of sustainable societal development.
- LEARN (and keep learning) and understand how to achieve social and behavioral change in practice.

- EXPERIMENT on how to actually implement the necessary changes and practical approaches ("trial and error").
- ENGAGE with multiple levers for change, including all individual, community, organizational, and government opportunities. Everything that is possible must be put to the test, whether it ends up being useful or not.

Previously, the International Energy Agency's 2021 report "Net Zero by 2050: A Roadmap for the Global Energy Sector" had pointed to the importance of change for everyone to combat climate change: "Behavioral changes play an important part in reducing energy demand and emissions... the Covid-19 pandemic has... demonstrated that people can make behavioral changes at significant speed and scale if they understand the changes to be justified, and that it is necessary for governments to explain convincingly and to provide clear guidance about what changes are needed and why they are needed."

The coming changes will be massive, and our collective consumption of everything will have to decrease significantly. Technology optimism may be good, but it will not help us much. Electric vehicles will not save us either. Our own carbon footprint per person will have to be reduced by two-thirds to roughly one-third of what it is today. Down to one-third! The problem right now is that low-carbon vehicles are still very expensive. An old gas-guzzler is much cheaper to buy than a new electric vehicle. A delicious vegan burger is still much more expensive than the beef burger from the fastfood spot down the street. At the moment, it is simply still too costly, difficult, and complicated for most people to switch to low-carbon products.

Ultimately, however, we must also see the positive side of this necessary change. A healthier, more resilient world will lead to people loving their homes more and being happier. Well-being and the improvement of our shared quality of life are things we should strive for as a global society. With that goal in mind, let us say goodbye to the good old days of grilling meat and being irresponsible to the environment and the general craze for faster, farther, bigger, taller. We should feel a deep sense of responsibility for ourselves, for our choices, and for our planet.

Greed and stupidity are the great global dangers of today — the greed to grab everything possible, to earn more, to steal more, and to build more. Growth feels like a societal imperative, but we need to discover something else — that not growing is part of the solution. The biggest threat to us is ourselves. Perhaps an -ism will take over US democracy, which is not a very far-fetched notion. Democracy in the US seems to be little more than a hollow shell after decades of internal erosion. "Election fraud" is just a pretextual accusation. Not trusting the state is the greater danger. The Ukraine war has shown that evil is still very much alive and has plenty of supporters. Greed, unrestrained growth, hate, fraud, and distrust must be fought, and above all, we must fight our own ignorance.

What will the future bring? There will be more of everything: an increasing number of storms, heat waves, disasters, and wildfires, and more system failures, challenges, and shocks. Each of these events will follow the other, or multiple collapses will happen at the same time. We will have to cope with all of this. We need to live more consciously, become more resilient, and be genuinely united. If we look at the coming disasters and challenges as a shared global problem, we can respond appropriately to the challenges with our large and flexible networks. However, we will never know what disaster is coming, so we should be well prepared for anything!

References

Academy of Social Science. (n.d.). Tackling climate change requires profound societal transformation. https://acss.org.uk/tackling-climate-change-requires-profound-societal-transformation/

Centre for Climate Change and Social Transformations. https://cast.ac.uk/

Centre for Climate Change and Social Transformations. (2022). Net zero literature review. September. https://our2050.world/wp-content/uploads/2022/09/Net-Zero-Literature-Review.pdf

European Science-Media Hub. (2021). Saving the planet: How far are we prepared to go? November 24. https://sciencemediahub.eu/2021/11/24/saving-the-planet-how-far-are-we-prepared-to-go/

Food and Agricultural Organization. (n.d.). Net Zero by 2050 — A roadmap for the global energy sector. https://www.fao.org/forestry/energy/catalogue/search/detail/en/c/1400699/

Fragkos, P. (2022). Decarbonizing the international shipping and aviation sectors. *Greece Energies*, **15**(24): 9650. https://www.mdpi.com/1996-1073/15/24/9650/htm

International Energy Agency. (n.d.). Net Zero by 2050: A roadmap for the global energy sector. https://iea.blob.core.windows.net/assets/7ebafc81-74ed-412b-9c60-5cc32c8396e4/NetZeroby2050-ARoadmap-fortheGlobalEnergySector-SummaryforPolicyMakers_CORR.pdf

Nielsen, K.S., Clayton, S., Stern, P.C., *et al.* (2020). How psychology can help limit climate change. *American Psychologist*, **76**(1).

Prosser, A.M.B. and Whitmarsh, L. (2022). Net Zero: A review of public attitudes & behaviours.

United Nations Educational, Scientific and Cultural Organization. (2023). Climate change education for social transformation. April 26. https://www.unesco.org/en/education-sustainable-development/cce-social-transformation

University of Bath. (2021). Social sciences to play vital role in meeting UK's net zero goals. Press release, February 1. https://www.bath.ac.uk/announcements/social-sciences-to-play-vital-role-in-meeting-uks-net-zero-goals/

Whitmarsh, L.E. (2005). A study of public understanding of and response to climate change in the south of England. Ph.D. thesis, April. https://doc.ukdataservice.ac.uk/doc/5345/mrdoc/pdf/5345thesis.pdf

Change — But How?

The real problem of humanity is the following: we have Paleolithic emotions, medieval institutions, and god-like technology.

— Edward O. Wilson (1929–2021), biologist, "father of biodiversity."

C hanging is one of the most difficult things to do. Once you learn how something is done, how it is done by others, or how it is done in your family, neighborhood, community, country, or culture, you not only get used to it, but you consider the habit or behavior to be correct, or even the only way to do it. Changing a learned behavior is quite difficult, even when it comes to the way you drink your tea or coffee. When it comes to climate change, many behaviors we have come to love must be discarded as "habitual garbage." The first thing we need to change is our relationship with each other. The most important factor in that matter is respect.

What does the process of change mean, especially in the case of climate change? It is significantly more difficult than simply switching gears on the highway. We are not cars, and we cannot change our minds by pushing a gear stick. Fortunately, experts have a lot to say about the process of change.

Change usually begins with ignorance or denial of a given problem. Either one is not aware of the problem at all, or one does not want to acknowledge it. At this stage of forethought, change is usually not perceived at all, and one's own behavior is similarly not seen as a problem. In the case of climate change, virtually all scientists (except those whose research has been or is being paid for by the fossil fuel industry) say that climate change is real and that it is caused by humans. If you will not or cannot acknowledge that, your only options are denial or ignorance — or perhaps both. If you find yourself at this stage, you should ask yourself some questions: Have I experienced something like this before? Under what circumstances and what occurred? Have I ever thought that I might have a problem? What would have to happen for me to recognize myself as the problem?

Then comes the phase of contemplation. You realize that there might be a problem. You do not quite know what to make of it, and you feel uncertain. You have started to think about a change, and you may have even weighed the pros and cons of a possible change. In the process, you encounter many obstacles and opportunities for escape. You start thinking about the cost and the fact that there will likely be some effort required as well. All of this is too much for you. This phase can drag on for a long time because many people are unable or unwilling to decide (at least this feels like a widespread phenomenon). It can take years, and some people never get past this stage, as the sacrifices they make emotionally, mentally, or physically outweigh the potential positive effects or gains of change. If you are in this stage of contemplation, you should ask yourself the following questions: Why do I want change in the first place? What is holding me back from change? What (event, situation, cause, conversation, incentive, impetus) would be necessary for me to change?

The next step in the whole process is the preparation phase. This is the phase where you prepare for a major change in your life, either something now or in the near future. For example, if you want to lose weight, you will start buying different foods for a low-carb diet. If you have to drive to work, you will find out about public transportation and the benefits of selling your car (it is actually much cheaper not to have a car). Think about the beauty of your country and how much you would enjoy a train ride, and perhaps you will choose to go upcountry instead of taking a long-haul flight to an overcrowded tourist trap where you will only find hip people and no peace of mind.

When you have made it to this point, you need to think about what would help you make a real switch. Now you need to gather more information, read books, and surf the Internet. Write down your motivation for this particular change. Write down a detailed plan with all your goals, intentions, and motivations — as well as how you will do it. Write down your own plan of action, then find support groups or good friends who can encourage you on your journey into new territory.

Be careful, because at this stage you really need to become active. You may realize that the path you are taking leads to a dead end. Sometimes, you will have failures because what you originally wanted to achieve is not turning out the way you imagined. Maybe you have started in the wrong direction and are doing the wrong thing — something that does not suit you. You need to learn that it can be much more difficult to accomplish what you planned at the beginning of your trip. The tour you planned might be too steep, the trip too long, and the compartment you are in too cramped. Or the vehicle you were relying on might let you down. Or the people you planned your trip with are just... not good for you.

Now you need to think about how to avoid setbacks and then start thinking about the really good things in your new life. Once you have sold your car, you will not have to worry about the price of gas, parking fees, the lousy auto repair shop, or accidents. The trick is that... you have to start! The best way to keep up with your journey is to reward yourself for making steady progress. You are the lucky one because you are moving forward! This thought will help you maintain your good mood for the next steps.

There will be moments that will feel like a break or even an interruption to you. This is your time in the "maintenance phase." Even if you do not feel it, you are still moving forward. Maybe not as fast as you thought you would or as you previously had, but when you are in the middle of the process, you will not feel that you are changing. The most important thing to remember during this phase is to avoid falling back into old habits and behaviors. Rather, stick to the new habits and immerse yourself in them. Enjoy your new surroundings, your new people, and your new life. Find out what it feels like in your new neighborhood — take a deep breath and relax. Let it sink in!

Just look ahead and do not turn around, lest you be tempted to fall back into old patterns. Instead, replace them with new activities and pleasures! When you avoid relapse, reward yourself immediately. Do not give up just because you feel like you have to, but remind yourself that it may just be a minor setback or short break. After all, you have already completed the crucial part of your journey into new territory. Reassure yourself of the positive impact your change efforts have had so far in a larger context. Look at yourself and start thinking about your next steps!

At this point in your journey, you might also have a feeling of deep disappointment. You might think that the whole journey has been absolutely useless. You are certain that you have failed or that the whole thing was and is pointless. You might be totally frustrated or painfully depressed. Now is the time to calm down. It is imperative that you re-evaluate your motivation and commitment to your goals. Remember what you had in mind when you started your journey. You knew (or were told multiple times by people who know) that change truly is the hardest thing to do and that there will be many things and many people seemingly set against you.

Tolerate them. It is their fault. You are on the right path, and relapses and disappointments are just normal. Remember, in his search for the right material for the filament of a working light bulb, Thomas Edison tested more than 6,000 different possible materials before he found one that fit the bill — charred bamboo. Maybe that number is too high, and maybe it is just a myth of his own making — and there were obviously other materials that came along later — but you get the idea: Never give up, especially when it comes to climate change and a better, healthier world.

References

Davis, T. (n.d.). How to change: 6 science-based tips & strategies. Berkeley Well-being Institute. https://www.berkeleywellbeing.com/how-to-change.html

Katy Milkman. (n.d.). How to change? https://www.katymilkman.com/book

Law, T.J. (2023). 9 powerful steps for how to change your life before 2023. Oberlo, June 10. https://www.oberlo.com/blog/how-to-change-your-life

Lestari. (n.d.). Education at a time of emergency. https://lestari.org/case-studies/education-at-a-time-of-emergency/

Our Paleolithic Minds

We live in a world where reality seems to revolve around us faster and faster. This phenomenon of our time makes many of us sick, or at least confused. The rising standard of technology and the rapidly growing digital infrastructures of postmodern reality are increasingly overtaxing the natural capacities of our brains. We live in a quasi-utopian world in which cultural and social boundaries are dissolving. Our world is no longer just the neighborhood next door, nor is it about the family matters we like or dislike discussing. Instead, we are experiencing a world where everything and anything exists together in one big mix. Recognizing overarching structures and keeping track of them has become very difficult for our overly simplistic brains and old thought patterns. We are analog people in a digital world of bits and bytes, where anything is possible. Extremism and conspiracy theories of all kinds flourish in the global playground of social media, without human editors to decide what is fact and what is fiction. This is sold to us as "freedom," and yes, we do have freedom of choice. We have to decide for ourselves, hundreds of times a day, what is right and what is wrong. However, this is becoming increasingly difficult or even impossible for too many of us because we lack the capacity, the knowledge, and the time.

Constant multitasking has become the new normal, but this makes our brains busier and more strained than ever before. We are bombarded with ever-increasing amounts of data, advertising, so-called information, TV, streaming, and other services. There is a constant stream of news and a barrage of fake news, facts and alternative facts, and gossip and rumors — all passed off as true. In order to find out what actual truth there really is in this flood of information, our brains are working in overdrive mode all the time, without interruption. The Internet world has not brought us all together in one place; on the contrary, we are increasingly alone and on our own in this world. We have to order our own food and tickets, book our own trips, and write our own contracts. We have to constantly download, install, learn new programs, and assimilate new updates in an ever-faster spinning wheel of tasks. Unfortunately, we have to do it, or else we will be obsolete and soon feel like we are living on the dark side of the moon.

Our phones are constantly reminding us of what to do and where to look. We no longer pay attention to the traffic or the people around us; we neglect the street life as we walk while writing text messages that may be superfluous or even not worth reading. We check our various email accounts while standing in line or having lunch with our dear friends, but do not worry; they are doing the same thing because it is the right thing to do if we want to keep up with the times. Aside from our bravado of being really hip despite (or because of) our grotesque multitasking efforts, we cannot keep up with it all. At the end of the day, we get tired of it, and our digital masters will relentlessly remind us of what we forgot to do. We are like donkeys constantly chasing a carrot. Today, the Internet Protocol has taken on the role of an embodied "Big Brother," and we must finally say goodbye to this scourge — not forever, not always, but for our times when we can become ourselves again.

We naturally try to meet the increasing demands of the modern world, but our brains remain unsatisfied. We simply cannot do enough to meet all the demands. We are not the people we see in advertisements, not the happy, incredibly rich, and completely satisfied people who lounge at the top of society — the healthy, successful, young people. Frustration, sadness, and maybe even illness seem inevitable. We feel miserable sometimes and maybe even all the time. We cannot be where society's norms want us to be. Earl Miller, an MIT neuroscientist, commented on this modern phenomenon: "When people think they're multitasking, they're actually just switching from one task to another very rapidly. And every time they do, there's a cognitive cost." We think we are doing everything right, but multitasking means stress, which increases the production of the stress hormone cortisol and the hormone adrenaline, which together can lead to mental confusion and tremendous anxiety.

We lose focus because we think we need to do more, and then we become addicted to the dopamine-fueled state of our brain during the daily stress of the multitasking flywheel. The more we do, the more we become addicted to doing more and more. We are on the verge of breaking down or burning out. Instead of focusing on a worthwhile task, we probably did many things we *thought* were necessary, but they left us unrewarded, unsatisfied, and lonely. We followed the predetermined path of multitasking and thought we did it well, but this is a deception created by our doped brains. In *Brave New World*, a novel by Aldous Huxley, the drug "soma" is freely distributed to people to make them feel good. Today, we no longer need such drugs; we live in a world that allows us to manufacture them with our own Paleolithic brains, just enough to make us feel somewhat normal.

Our Paleolithic brains are not (yet) made for holistic consciousness. We need boundaries and rules and time to think. The daily overdose of information and the stress and horror we are exposed to every day put our brains in a constant state of helplessness. There is no soothing voice to coddle our hurt feelings, no guiding light in a world of agonizing information streams devoid of any truth, and no theory to translate into everyday rules for everyone. We are staring down a seemingly endless rabbit hole with no light at the end of the tunnel.

The dissolution of the world has arrived within us. Sometimes people feel the emptiness. This situation is causing the already existing anxiety of many of us to rise beyond a tolerable level. There is only one way out: trust in ourselves, in our own intellect, and in our courage. However, since our Paleolithic, analog minds are not made to find new convictions in a digital world or to satisfy and calm our hungry brains, we are driven toward more and more action and consumption. In a world based on growth and constant movement for all, rest is not allowed, which leaves all of us with an uncertain feeling in today's unforgiving world. We would like to believe that everything is fine, but it is not. We have to find a way to adapt to these new conditions of life, even as the world spins faster and faster. Now, with a looming catastrophe on the horizon, a growing level of fear is leading to more and more pessimism, which could become a deadly force of its own, destroying any hope among humanity that remains. The COVID-19 pandemic has shown at least one thing: Decelerating the spinning wheel of a society built entirely on futuristic new technologies is possible.

References

BRAND MINDS. (n.d.). To multitask or not? This is the question. https://brandminds.com/to-multi-task-or-not-this-is-the-question/

EOH. (2019). Technology is overpowering our Paleolithic brains. September 26. https://www.eoh.co.za/technology-is-overpowering-our-paleolithic-brains/

Harris, T. (2019). Our brains are no match for our technology. *The New York Times*, December 5. https://www.nytimes.com/2019/12/05/opinion/digital-technology-brain.html

Miller, E.K. (n.d.). Multitasking: Why your brain can't do it and what you should do about it. The Picower Institute for Learning and Memory and Department of Brain and Cognitive Sciences, MIT. https://radius.mit.edu/sites/default/files/images/Miller%20Multitasking%202017.pdf

The Change

The difficulty lies, not in the new ideas, but in escaping from the old ones, which ramify, for those brought up as most of us have been, into every corner of our minds.

— John Maynard Keynes,
The General Theory of Employment, Interest, and Money, 1936

Psychologist Brian Jeffrey ("BJ") Fogg of the Persuasive Technology Lab at Stanford University in California has found a relatively safe way to trigger change in people. According to his model, three elements must occur simultaneously: motivation, ability, and triggers (or prompts) so that people are induced to change their old behavior and embrace a new one.

Motivation (high to low): The more motivated you are to do something, the more likely you are to do it. There are six main motivators, which are divided into three categories:

- Sensation — Pleasure/Pain
- Anticipation — Hope/Fear
- Social Dimension — Social Acceptance/Rejection

Ability: (hard to do or easy to do) with six subcomponents:

- Time — The change should not take much time, or people will not be willing to do it.
- Money — If you cannot afford the behavior change, you will not do it.
- Brain Effort — Any new behavior should not increase the cognitive load too much.
- Physical Effort — The less physical effort required, the more likely you are to do it.

- Social Deviance — Few people will change their behavior if it violates social norms.
- Non-Routine — It is much easier to adopt a new behavior when you can incorporate it into your routine.

Trigger: (failure or success) A trigger is a call to action; it says, "Do this now." In the Fogg model, there are three types of triggers:

- Spark — Needed when you have high skills but low motivation. It should contain a motivational element.

  Example: You wake up lying in your warm bed. Why should you get up?

  Solution: Set a really loud alarm clock and place it quite far from your bed. When the alarm goes off, you will surely get up to get rid of the noise!

- Facilitator — Needed when motivation is high but skills are low. A facilitator tries to simplify the task.

  Example: You want to eat healthier but lack organization.

  Solution: Sign up for a newsletter that provides easy-to-prepare and delicious recipes for healthy meals. This will get you shopping, organizing, and planning your healthy meals.

- Signal — Used when both motivation and ability are high, but you need a clear reminder. The signal can even be just a simple yellow Post-it note.

If you want to change but are not sure you will, consider the following:

- Do I have a motivational problem? How can I fix it? Which of the six core motivators should I use?
- Do I have a capability problem? If so, how can I fix it? How can I make the behavior easier or simpler? What resource is needed most (time, money, physical ability, etc.)? How can I address the need?
- Am I responding to the right trigger? At the right time?

References

BJ Fogg, PhD. https://www.bjfogg.com/

Center for Humane Technology. (2021). Persuasive Technology: How does technology use design to influence my behavior? August 17. https://www.humanetech.com/youth/persuasive-technology

Fogg, B.J. (2009). A behavior model for persuasive design. Conference: Persuasive Technology, Fourth International Conference, PERSUASIVE 2009, Claremont, California, US, April 26–29. https://www.demenzemedicinagenerale.net/images/mens-sana/Captology_Fogg_Behavior_Model.pdf

Keynes, J.M. (1936). The general theory of employment, interest, and money. February. https://www.files.ethz.ch/isn/125515/1366_keynestheoryofemployment.pdf

Stanford University. (n.d.). Behavior design lab. https://behaviordesign.stanford.edu/

Wikipedia. (n.d.). https://en.wikipedia.org/wiki/B._J._Fogg

Climate Anxiety

C limate change is very scary. Being afraid is a very deep and dangerous feeling, but when people feel a threat, they are instinctively more motivated to take action. That is the advantage of fear — it makes people alert. The question is: How can we deal with these feelings of fear, stress, and anxiety about the threat of climate catastrophe, even as it descends upon us?

Social scientists have identified a wide range of coping strategies for dealing with climate change. Some of these strategies can be very helpful, such as taking a problem-focused, neutral stance, seeking social or personal help and support, or trying a realistic, active coping approach.

Here are some positive behavioral strategies that can be used to manage potentially distressing feelings of climate stress and anxiety:

- Become active! It is always the best solution to be proactive against your own fear. Review your carbon footprint and think about how you can reduce it. Work to change your life and your attitude. Turn down your heating, sell your car, ride your bike, take the train, and use public transport. And if you are brave enough, meet with climate protection organizations and become part of the movement.
- Take a break! Calm down, and do not worry too much, but do not give up on your climate awareness. Just try to relax, take a day or a week off from thinking about the climate crisis, and regain some momentum. The fight against climate change will take much longer than you think. Since this is a marathon, you need to stay alert and in good health!
- Remotivate yourself with positive experiences! Make sure you feel good; choose activities that evoke positive emotions of passion, joy, and fun! It

is your own responsibility to stay positive! Build a positive momentum that you can pass on to others, be kind to yourself, be passionate, and find people who help you develop good feelings.

- Routines are very important for a stable and healthy life. Make sure that your habits help you feel good. Exercise occasionally, including running, swimming, or yoga. Eat a healthy diet and get enough sleep. Go out and enjoy the natural world that you want to protect so you know what you want to preserve.

- Focus! Do not try to be involved in everything. Too many projects can be overwhelming, so focus on the things you really want to do and enjoy. You cannot be everywhere, so pick and choose where you invest your energy.

There are also relationship strategies that can help us deal with climate stress and anxiety. We are social creatures, so good and healthy relationships with others are the foundation for a fulfilling and balanced life. When we care for and support others, we are important, active members of the social web in which we live. A vibrant social life is one of the best ways to deal with stress and anxiety. This kind of support greatly reduces mental stress in times of global challenges. Talk to friends and colleagues to share your concerns about rapid climate change. Be or become an active member of your community and find out who shares your values. Find a mentor who can help you understand the issues around you — someone who can and will give you strategic advice. It will be a very satisfying experience to work on new projects for the future with a group of strong, like-minded people.

Cognitive strategies are very helpful in stressful situations. Use your mind to deal with stressful feelings! Under stress, many people can fall into a number of unhelpful thinking patterns, e.g., black-and-white thinking such as "driving is absolutely bad and should be punished" or overreactions such as "this apocalyptic rainfall shows that we are all going to die soon."

Do not fall into this trap, and instead, be careful to recognize such malicious patterns. Replace them with more useful ones and see the positive side of this process. To get yourself to think more realistically and even encouragingly about the problems, say something such as, "History has proven that we humans, meaning you and me, can change." Or, if your starting point is, "Unfortunately, climate change is happening much faster than expected, and that means there's nothing we can do about it," change that negative view to, "Scientists are certain that we still have a chance to cope with climate change and prevent the worst-case scenario — the end of the world."

The most important thing is to let go of your negative feelings and change them for the better. Do not call yourself an idiot, as we all tend to make mistakes. And do not blame yourself for not doing enough to create a better, healthier world. Just be more productive and turn those self-directed accusations into, "Next time, I'll be better prepared to solve the problem."

The next most important thing is that you take enough time to think. At that point, you can break down your long-term goals into smaller steps, which will make those goals much easier to achieve. Using these methods, you will also find better perspectives to see things through and how they will turn out. Nothing will feel so urgent that it will overwhelm your patience. Little by little, things take time to grow. Have the courage to take these small steps and then move on to solving more difficult tasks. Do not get discouraged. Stay hopeful. Pass this feeling on to others as well, and keep it up. Turn your anxieties into a source of your well-being and hopes; turn them into action. It will pay off in the end.

Find emotional strategies. Try to learn to control your feelings. Humans have the ability to change their feelings from "too anxious" to "not so anxious." Wait for the right moment to do this, and take the time to let it happen. It will come naturally. Name your feelings and address them.

They will respond to you and show you where to go. In this way, you can counteract the feeling of fear and use it as a motor or a battery to regain your essential strength.

References

Collier, S. (2022). If climate change keeps you up at night, here's how to cope. Harvard Health Publishing, June 13. https://www.health.harvard.edu/blog/is-climate-change-keeping-you-up-at-night-you-may-have-climate-anxiety-202206132761

Dodds, J. (2021). The psychology of climate anxiety. *The British Journal of Psychiatry*, **45**(4): 222–226.

Mertens, M. (2023). Why aren't we more scared of the climate crisis? It's complicated. *The Guardian*, July 22. https://www.theguardian.com/environment/2023/jul/22/climate-crisis-fear-psychology

Change Is a Constant

C hange is part of our daily lives. For our really big goal (creating a more resilient, sustainable, and greener world), we need to focus on a different form of change — a willful, purposeful change of mind. In doing so, we are first confronted with an old adage: "You can take man out of the Stone Age, but you can't take the Stone Age out of man." This is the short version of the core statement of evolutionary psychology, namely that man has retained the mentality of his Stone Age ancestors. This is based on scientific research proving that certain ancient patterns of behavior are repeated in all societies, regardless of time and place. These patterns are believed to have a biogenetic origin.

Scientists have identified several genes that are thought to control human predispositions, such as temperament and cognitive abilities. Paleontologists have discovered how humans lived very long ago and how their characteristics gradually adapted to the environment they inhabited. Social psychologists have studied the conditions under which people cooperate, compete, or behave aggressively. Their findings suggest that certain impulses and/or actions are hard-wired into the human psyche. On this basis, adherents of evolutionary psychology directly contradict the common management theory that says people can change if they are properly trained and/or motivated.

Certainly, change is a process of transformation that can unsettle people in threatening ways. Sometimes, change is as profound as a complete reinvention of yourself. You must bravely face the loss of things you leave behind and embrace the uncertain future that awaits you. And you have to be prepared for the moment of change. It tends to be a long process, but there will be moments when it "clicks," and you feel that something big has happened. You have to prepare for these moments, as they can be very moving. Sometimes, they are so tiny that you may miss them when they happen, so keep your eyes open. And remember, timing is everything, especially when it comes to change.

In organizations (companies), the 3-phase model for change is still current — breaking up the existing level, moving to the target level, and consolidating group life at the new level (Unfreezing – Moving – Freezing). "Unfreezing" means the overcoming of inertia accompanied by the dismantling of the existing mindset. "Moving" is the process of change or transition. "Freezing" is when the new mindset is in place, along with a new and high comfort level. This simple, three-stage model was purportedly developed by Kurt Lewin (German-American psychologist, 1890–1947), while recent research has found no evidence of its authorship. This simple process may work well in organizations, but for people, it may look different. There is an 8-step model for this (after Dr. John Kotter):

- Establish a sense of urgency
- Formulate a powerful mission statement
- Create a vision
- Communicate the vision
- Remove obstacles to enable action
- Plan and achieve short-term success
- Consolidate improvements and bring about further changes
- Institutionalize new approaches

Change starts small. Do not be afraid of the long journey ahead of you. It will start with one small step and go on to the next, then another. It is a journey from day to day, step by step. It might be a painful process because if we want to change, we have to say goodbye to old beliefs and grieve over them. Sometimes, the process will stall, and you will say, "I just can't move on. This was a bad idea from the beginning. I have to give it up." In that case, tell yourself that maybe you are just having a bad day. Remember, a bad day knowing you are on the right track is better than a good day of being on the wrong track.

Do not rush. Change is a process during which you have to convince yourself of the value of change. Long-term success is based on small steps in the right direction. Be patient and stay committed despite the inevitable ups and downs. Each step forward takes you further away from old dogmas and old patterns and further away from the mistaken belief that uniformity means security. Each step of your journey brings you closer to the world we all ultimately want — a healthy, sustainable, resilient, and friendly world for us all.

The science behind behavior change is complex. One institute dedicated to its study has a complex name too: Science of Behavior Change (SOBC) Resource and Coordinating Center (RCC) at the Columbia University Department of Medicine. Another, the Centre for Climate Change and Social Transformations (CAST), is something of a joint venture between Cardiff University, the University of East Anglia, the University of Manchester, the University of York, the University of Bath, and the think tank Climate Outreach. Both institutions are deeply involved in the science of climate change and are worthy of close consideration.

Climate Outreach is a registered charity in England and Wales. It has its own theory of change, which focuses on the practical question of how to get from where we are to where we need to be. That is a good starting point for social scientists and communications specialists who "help people understand this complex issue in ways that resonate with their sense of identity, values, and worldview. Informed consent and support from people across society and around the world creates what we call a social mandate for climate action — and we believe it is how real change happens."

Climate change requires proactive action around the world and at all levels of society. This planet-wide pivot cannot be achieved by doing things by the book. In fact, there are no books for the unprecedented situation of today, where a very rapid change in attitudes and social norms is our only chance for survival. Within a short time, new behaviors must be introduced throughout the population in different ways, in different places, and at different times. This requires extensive communication and the active engagement of the public, or it will not succeed.

As we know, significant emissions reductions are an essential part of the coming climate revolution. These reductions will require a change in lifestyle for everyone and a huge shift in how environmentally friendly behavior is recognized. Unfortunately, it seems that the necessary changes are not talked about enough; crudely put, this topic seems to be avoided by most people. Apparently — again — people do not talk about things that are unpleasant or undesirable. It echoes the case after World War II in Germany, when no one liked to talk about guilt or shame or even what some people had done during the reign of the Third Reich. The shortcut to getting rid of the bad feelings about the past was to say (and believe) that the only one who was really guilty was already dead — Adolf Hitler. And those who had committed atrocities and were responsible for the war itself, especially its outcome, were the Nazis. And since no one remaining had been a Nazi (it was always someone else), no one was responsible. It has taken more than a generation to lift the fog of silence, and the tendency to forget rather than face the facts is still widespread.

That is why it is first necessary to bring the discussion of radical climate action onto a broad public stage. Fridays for Future sends clear signals in this respect, but even this wake-up call has not yet led to a discussion across all levels and strata of society. To be successful in the long term, the commitment of the general public must be awakened. To break the silence and wake up the public, we first need to understand why it is so difficult to get people to talk about the upcoming changes.

- The main reason for rejecting the necessary steps to avoid climate catastrophe is that people pursue their own interests. They are in the fossil fuel industry, conventional agriculture, or the steel industry. Or they just want to hold tight to the old fossil fuel economy because that system has been working ideally so far, and they fear change of any kind.
- If the discussion is mainly led by left-wing groups, it will probably be even more difficult to achieve a broad discussion in all layers of society. Slowly, the impacts of climate change are being seen as eminent even by conservative groups. The middle of society, in particular, must acknowledge the challenges of climate change in order to become immune to the climate deniers. It could even be that the knowledge about climate change as such and about its causes will be negated, and other problems (economy, defense) will be brought to the fore in order to avoid the discussion about climate change altogether.
- The discussion about climate change is dominated by scientists or science-oriented celebrities and science-related language. This approach is largely ignored by ordinary people. They do not accept people who talk smart and clever. The whole issue of climate change and its impact on society needs to be framed in a different, more understandable, and simpler way. The discussion thus far has been too academic, and, therefore, too detached from people's real concerns.
- If people do not identify with the community fighting climate change, they will more or less neglect the problem. The entire movement will gain momentum only if it succeeds in inspiring every demographic and social stratum to protect the planet.
- People need to be convinced that their commitment pays off, especially for themselves. We all know at the bottom of our hearts that living in healthy and sustainable environments and communities is the better future in the end.

Change will not occur just because we are told what to do. Change comes about through interactions between people. People base their behavior on what those around them are doing and saying. Social proof is king. What matters is communication with each other — not what happens in the media. Climate change must become part of the everyday concerns of large groups in society. Concern about climate change must become our collective concern so that sustainable and committed action becomes possible.

The story of the tobacco industry can serve as a blueprint for how the story of climate change might unfold in the future. Tobacco was a cornerstone of society for decades until, over time, its impact on people's health became clear. In a US Congressional hearing, the CEOs of major tobacco companies were asked what they thought of the health risks of their product. They all wholeheartedly denied any risk ("Nicotine is not addictive!"). Perhaps a similar story will play out with today's fossil fuel industry CEOs. Only this time, the problem is much bigger because, this time, we are all at risk.

References

Climate Outreach. https://climateoutreach.org/

Columbia University. (n.d.). Science of behavior change. Center for Behavioral Cardiovascular Health, Irving Medical Center. https://www.columbiacardiology.org/research/research-centers-and-programs/center-behavioral-cardiovascular-health/research/science

Columbia University. (2018). The science behind behavior change. Irving Medical Center, February 20. https://www.cuimc.columbia.edu/news/science-behind-behavior-change

Forbes Councils. (2022). The 6 best change management models for your company. December 22. https://councils.forbes.com/blog/top-change-management-models

National Institutes of Health Common Fund. (n.d.). Science of Behavior Change (SOBC). https://commonfund.nih.gov/behaviorchange

Science of Behavioral Change. (n.d.). Resource and Coordinating Center. https://scienceofbehaviorchange.org/projects/resource-and-coordinating-center/

Study.com. (n.d.). Lewin's 3-stage model of change: Unfreezing, changing & refreezing. https://study.com/academy/lesson/lewins-3-stage-model-of-change-unfreezing-changing-refreezing.html

Understanding Global Change. (n.d.). Evolution. https://ugc.berkeley.edu/background-content/evolution/

Wikipedia. (n.d.). Change management. https://en.wikipedia.org/wiki/Change_management

Wikipedia. (n.d.). Evolution. https://en.wikipedia.org/wiki/Evolution

YouTube. (2012). 1994 — Tobacco company CEOs testify before Congress. July 4. https://www.youtube.com/watch?v=e_ZDQKq2F08

The American Way

K aty Milkman is an American economist and author of the book *How to Change.* Her advice for successful change is to start with a blank sheet so that you can free yourself from your old life and start a new one, or at least a new chapter or episode. To really free yourself and turn to the new with strength and conviction, you should also find a good title for your new life. For example, "My New Life," "The New Me," or "The Green Age."

You should prepare wisely for a sustainable life; that is, you must first learn to strengthen your inner self for the coming struggle. In the future, you will face many more storms, violent floods, tornadoes, and the like. These might not necessarily happen in your area or country, but climate scientists are certain that these disasters will definitely become more frequent. Though it cannot be said exactly where they will occur, they will happen, and they will have an impact on your well-being. As a result, you will experience conflict, costs, and chaos, but you should respond to it as calmly and appropriately as possible. Generally, there are three ways we can prepare for and respond to climate change: Mitigation, Adaptation, or Suffering. Mitigation means trying to help stop climate change, while adaptation means coping with a changing climate. Please try to avoid the last one.

You must get through everything that is coming, so you have to remain stable in health and mind. Calmness and composure will not only help you to cope with climate change but also to meet other people in a nice and friendly manner. Do not spread unnecessary stress! This strategy will not only help you adapt to the changing climatic conditions but also remain relaxed and confident. Others may struggle to do this, as they have not started their own change at all.

Do not be fooled by fake friends or self-proclaimed experts who tell you that "it will all go away." It will not. There will be more and more climate refugees around the world in the near future. It could very likely happen in your country. You should also be prepared for such changes — possibly in your own neighborhood — and be ready to help others. We need much more cohesion in the future because no place in this world is safe from climate change.

You also need to be prepared for surprises and possible emergencies. You should know where to go for safety in case of emergencies (e.g., heavy rains with flash floods). You should also adjust your clothing — practical clothing will be the new chic. Various bugs, ticks, and mosquitoes have already followed climate change, so good footwear and insect repellents are already essential in many forest areas. Find out who is familiar with these novel requirements in your area; there is almost certainly helpful information online. Pass this valuable information on, as it will be useful to others as well.

However, the most important thing on your journey is to be confident and to stay in a good mood. Be prepared for the unprepared people because they are the ones who will need your help in the event of an emergency. Do not become pessimistic or even depressed but remain optimistic. Work hard and with a conviction to feel good about yourself even if a disaster happens. Most importantly, keep changing because the world is changing rapidly and in unpredictable ways. Perhaps more importantly, always remember that everything necessary for change already exists. We just need to find new ways, new technologies, and new approaches. In the end, everything will be fine.

David Pogue, author of *How to Prepare for Climate Change*, offers practical guidance on adaptation actions you can take right now. The best advice he gives in his book is to act quickly. Quite simply, it is time to change and to change as radically and as quickly as possible.

References

Bradford Council. (n.d.). How to prepare for climate change. https://www.bradford.gov.uk/environment/climate-change/how-to-prepare-for-climate-change/

Climate-Adapt. https://climate-adapt.eea.europa.eu/

European Commission. (2021). Forging a climate-resilient Europe — the new EU Strategy on adaptation to climate change. February 24. https://eur-lex.europa.eu/legal-content/EN/TXT/?uri=COM:2021:82:FIN

Madill, E. (2016). 11 Powerful Tips to Keep Your Spirits Bright. *Huffpost*, March 31. https://www.huffpost.com/entry/11-powerful-tips-to-keep-your-spirits-bright_b_9570728

Patte, E. (2022). How to prepare for climate change's most immediate impacts. January 16. https://www.thegrowthfaculty.com/blog/howtochange

Semph, C. (n.d.). 10 ways to lift your spirits when you're having a tough week. Tiny Buddha. https://tinybuddha.com/blog/10-ways-lift-spirits-youre-tough-week/

Suarez, I. (2020). 5 strategies that achieve climate mitigation and adaptation simultaneously. World Resources Institute. February 10. https://www.wri.org/insights/5-strategies-achieve-climate-mitigation-and-adaptation-simultaneously

The Growth Faculty. (2022). Katy Milkman: How to change despite those obstacles inside you. August 31. https://www.thegrowthfaculty.com/blog/howtochange

United States Environmental Protection Agency. (n.d.). Strategies for climate change adaptation. https://www.epa.gov/arc-x/strategies-climate-change-adaptation

Verplanken, B. and Roy, D. (2014). HABiT (Habits, attitudes, and behaviours in transition). Sustainable Lifestyles Research Group. June 3. http://www.sustainablelifestyles.ac.uk/projects/change-processes/habits.html

Wikipedia. (n.d.). Katy Milkman. https://en.wikipedia.org/wiki/Katy_Milkman

What Is Liberty?

The limits of liberty were defined in the French "Declaration of the Rights of Man and of the Citizen" of August 26, 1789, Article 4: "Liberty consists in being able to do anything that does not harm others: thus, the exercise of the natural rights of every man has no bounds other than those that ensure to the other members of society the enjoyment of these same rights. These bounds may be determined only by Law."

In 1793, Immanuel Kant summarized the limits of the legal liberty of man in an enlightened constitutional state (from: "On the Common Saying"): "No one has a right to compel me to be happy in the peculiar way in which he may think of the well-being of other men; but everyone is entitled to seek his own happiness in the way that seems to him best, if it does not infringe the liberty of others in striving after a similar end for themselves when their Liberty is capable of consisting with the Right of Liberty in all others according to possible universal laws." All these philosophical acrobatics have been simplified in a modern way to: "The freedom of one ends where the freedom of the other begins."

In pop folklore, it became "Freedom's just another word for nothin' left to lose," a line from singer-songwriter Janis Joplin's song *Me and Bobby McGee*. The singer, who died at an early age, gave the booze-rock-hippie culture of the American West Coast a trademark that was probably right at the time. Today, freedom is just about the last thing anyone wants to lose, in view of the millions of vaccination opponents who pit our freedom to be healthy and protected against their own freedom of speech or in view of the numerically even larger group of conspiracy theorists who take the liberty to talk nonsense and even believe it. Or take the many Putin believers, not only the millions in Russia but worldwide. Freedom must always be defended because it is a most precious thing, and it always and everywhere has powerful opponents.

Freedom is always dependent on external influences (e.g., the environment) and on other people. We cannot and should not pretend that freedom means that you can do anything. The motto is that freedom must be appropriate to the circumstances. Freedom does not mean that it is right and proper to just have fun and not bear any responsibility. Today, enlightened persons should be able to look beyond themselves to recognize and respect the person beside them as well as their needs. A general mindfulness must be found, cultivated, and preserved. Only this can be the basis of our freedoms and liberties in the future.

Climate change is an existential threat on a planetary scale. Responding to this crisis requires reforms that can only be undertaken collectively by all the world's governments. Individuals can, at most, be responsible for their own behavior, but governments have the power to enact laws that force industry and individuals to act sustainably. Although the power of consumers is strong, it is dwarfed by international corporations, and only governments are in a position to regulate companies and entire industries in such a way that the interests of citizens to live in a sustainable environment are safeguarded.

The responsibility of the individual toward the community is a fundamental aspect of human coexistence. In the crisis we face today, rapid, far-reaching, and unprecedented changes are needed in all areas of society, according to the 2021 report of the UN's Intergovernmental Panel on Climate Change (IPCC). Just how rapidly things have actually changed in recent decades is demonstrated by the fact that more CO_2 has been blown into the atmosphere since 1988 than in all of human history to that point. We are on a determined path to suffocation and climate collapse if we continue as we are.

People need to shift their lifestyles and consumption patterns to more sustainable alternatives, especially in areas they can control, such as transportation, buildings, and food habits. To reach the 1.5°C target, people need to travel less, and, in general, consumer preferences need to shift toward more sustainable choices, such as replacing private cars and air travel with public transit, buses, and trains. Buildings and cities need to become smarter, building materials need to shift from concrete to other materials such as wood and bamboo, and in the future, we need to heat our lives with renewable resources rather than fossil fuels. People need to consume fewer animal products, as the livestock sector is estimated to account for approximately 14.5% of global greenhouse gas emissions.

These so-called Shared Socioeconomic Pathways (SSPs) describe possible development trajectories and lend new narratives and lines of development to climate modeling. The need to change is a good thing; it will pay off for everyone, as all these changes mean more chances and opportunities for a modern, enlightened, and healthy world. We all have the freedom to decide what we want to do, but we must choose the path we want to take wisely.

Evolution shows that climate disruption is always associated with the widespread demise of life forms. This is exactly the process we are going through, but we have the freedom of choice between extinction and enlightenment. This is our responsibility that we must face. If we want to return to the path of sanity, we must end certain types of technologies. Only from their ashes will new behaviors and manifestations emerge.

Psychological effects related to climate change include the loss of traditions, habitat, and cultural heritage, as well as the despair that comes with having to leave the land where your ancestors are buried and where you have spent your whole life. All of this creates fear and terror. Smoke-filled skies, wildfires caused by prolonged drought, or another pandemic may elicit sadness, depression, and hopelessness. While we humans are incredibly adaptable, we are also enormously vulnerable to mental disorders.

Mitigating climate change will be difficult, and it will not work without individual sacrifice. We must do everything we can to avoid millions of needless deaths. We are already on the brink of a possible international confrontation and are already in a covert Second Cold War, but we must fight through to ensure our survival. Let us do it while we still can.

It's never too late to be what you might have been.

— George Eliot (1819–1880), English writer

References

Cambridge Dictionary. Liberty. https://dictionary.cambridge.org/dictionary/english/liberty

Élysée. (n.d.). The declaration of the rights of man and of the citizen. https://www.elysee.fr/en/french-presidency/the-declaration-of-the-rights-of-man-and-of-the-citizen

Hausfather, Z. (2018). Explainer: How 'Shared Socioeconomic Pathways' explore future climate change. Carbon Brief, 8 April. https://www.carbonbrief.org/explainer-how-shared-socioeconomic-pathways-explore-future-climate-change/

Intergovernmental Panel on Climate Change. (2022). Climate Change 2021: The Physical Science Basis. Working Group I contribution to the sixth assessment report of the Intergovernmental Panel on Climate Change. https://www.ipcc.ch/report/ar6/wg1/

Kant, I. (1784). Kant's principles of politics, including his essay on perpetual peace. https://oll.libertyfund.org/title/hastie-kant-s-principles-of-politics-including-his-essay-on-perpetual-peace

Larbi, R.M. (2020). The freedom of one ends where the freedom of the other begins. Atalayar, October 23. https://www.atalayar.com/en/opinion/ramdan-mesaud-larbi/freedom-one-ends-where-freedom-other-begins/20201023104652134884.html

Wikipedia. (n.d.). Liberty. https://en.wikipedia.org/wiki/Liberty

Wikipedia. (n.d.). Shared socioeconomic pathways. https://en.wikipedia.org/wiki/Shared_Socioeconomic_Pathways

Moments

E vents that break existing patterns or routines provide wonderful, unpredictable opportunities for lasting change. Examples of these moments of change, when place and time can open up to new dimensions, include:

- "Catharsis," a sudden transformation process that paves the way for fundamental personal or psychological change.
- "Rites of passage" that mark deep and lasting transition processes in people's lives.

These moments of change are also called entry or turning points, fateful moments, or transformative moments. Everyone has experienced such moments. We all know what these moments mean, as well as how powerful and valuable they are in life. These moments open the gates to something new, and they offer opportunities to truly and effectively change our behavior. They are catalysts for something that was in the making or had been planned for a long time but never happened. We should take advantage of these moments because they effectively end old habits and create space for new experiences. We all need to go through these gates to enter new territory.

People are currently being born into a postmodern society run by fossil fuels. Governments and the (financial) industry must now take the lead to transform the fossil oil and gas industry through legislative action. What we need is an ecosystem based on individuals willing to change their own behavior and elect representatives who will work for a thorough transformation of society. We also need governments that are willing to fight for their citizens, not for the fossil fuel industry. Thus, it takes educated citizens on the one hand and responsible politicians and governments on the other.

We need to change our economic, financial, and societal systems. Our ecological system is based on nature, and nature has already shown us the results of our economic system: droughts, floods, forest fires, and more. It only took our capitalist system four centuries to destroy a system that has been evolving for 4 billion years. This brutal, horrendous discrepancy shows how quickly we need to change our system. Perhaps we need to start questioning what has led us to think that this world and our greedy hunger for more could be endless. This question must be asked again and again until we find the right answer.

References

CE Change (n.d.). Responsible change: How do we deliver transformation responsibly? https://www.cechange.co.uk/responsible-change-how-do-we-transform-people-processes-and-performance-responsibly/

Centre for Climate Change and Social Transformations. (n.d.). Welcome to the CAST data portal. https://cast.ac.uk/publications/cast-data-portal/

The Garrison Institute. (2017). The responsibility to change. December 14. https://www.garrisoninstitute.org/blog/the-responsibility-to-change/

Tyndall Centre For Climate Change Research. https://www.uea.ac.uk/climate/tyndall-centre-for-climate-change-research

The Lucky Chance

I n 1754, the British historian and writer Horace Walpole (1717–1797) wrote to a friend about an unexpected discovery. The discovery concerned a painting by Giorgio Vasari, presumably lost, which referred to the Persian fairy tale "The Three Princes of Serendip." The title of the painting is derived from the ancient name Sri Lanka, which comes from Sanskrit. The princes depicted in the painting were known for repeatedly making accidental discoveries. Serendipity, according to Webster's Dictionary, is "the faculty or phenomenon of finding valuable or agreeable things not sought for," or according to Encyclopedia Britannica, "luck that takes the form of finding valuable or pleasant things that are not looked for."

Serendipity might be a very good method for coping with multi-faceted problems such as climate change. For example, if you misplace your car key yet again, you already know how hard it is to look for it under pressure. Therefore, do not think about it, and you will find the key. This is how the "serendipity method" works: do not concentrate, but be open to everything. How good this method is can be argued for by the fact that several valuable inventions that we all know were made with this "non-method" — the microwave oven, penicillin, the popsicle, and the Post-It note.

Legend has it that Elon Musk, the entrepreneur behind Tesla and Space-X, works on a similar principle — what he calls "first principles thinking." Musk basically questions every assumption he thinks he "knows" about a particular problem. For example, Musk was thinking about how to lower the cost of space flights. He came up with the idea of reusing rockets rather than just using them once. The principle was tested and failed several times before it finally worked. Today, his company, Space-X, is a NASA contractor planning trips to Mars.

Thomas Edison, the model for almost all great entrepreneurial inventors of the modern world, is another example of this principle. Edison, however, developed his very own technique. When he ran out of ideas, he would sit down on a chair, a heavy ball in his hand. When he started to nap, i.e., when he was halfway to sleep ("hypnagogia"), he let go of the ball and was awakened. Then he continued with his work, faster and better than before. This method of coming up with ideas while half-asleep was recently confirmed in a study by the University of Paris to enhance creativity. Edison focused not on what he wanted to invent but on the fact that he wanted to invent something. And he did not just want something small but always something that could change the world. He opened his mind to it, and he worked until he found something revolutionary.

The first thing to do (according to Edison) is to concentrate deeply. The second rule of Edison's path to success is to let go and put yourself in a semi-slumber, with patience and willingness to learn during the process. Failure in this process is not a bad thing but something necessary to obtain better results. Overall, Edison's method was profoundly good. At the end of his life, Edison had 2,332 patents registered in his name.

The serendipity principle relies on radical, out-of-the-box thinking, perhaps even thinking that goes against all the "established" rules. People who adopt this way of thinking and finding will meet resistance or even rejection. Probably, they might not initially be loved because they do not explicitly follow rules and thus create friction around them. People who walk this path will have to be brave, but they will be rewarded with discoveries and perhaps even social recognition. Serendipity will not just happen to you;

it will happen *because* of you. Serendipity does not happen by accident but because you take chances, take action, and connect with people. You follow a path, and eventually — after many trials and failures — you find something that you were not looking for. But you do find something truly innovative that may be great for the world.

Figure 49.   Mural of the Wheel of Life (Trongsa Dzong) in Bhutan. (Source: Von Stephen Shephard, CC BY-SA 3.0, https://commons.wikimedia.org/w/index. php?curid=1130661)

The Tibetan word "tendrel" is the translation of a Sanskrit term "pratitya-samutpada," which can be translated as "dependent causality, interdependence, relativity, or serendipity." The term represents the Buddha's teaching on cause and effect, and it shows how all situations in our lives arise from the interaction of various factors that are interrelated. In Hinayana (the main form of Buddhism in Southeast Asia), the term refers specifically to the 12 *nidanas* (Sanskrit for cause, occasion) depicted in the Wheel of Life diagram. Only when the practicing Buddhist succeeds in breaking one of these connections — for example, by overcoming ignorance or addiction — does the entire cycle of suffering cease, and one attains enlightenment.

Figure 50.   Rose "Serendipity." (Public Domain)

How can the serendipity principle be integrated into our daily lives? First, you need a "prepared mind." Artists do this all the time; it is actually a key part of their creative process. You must establish a studio, a writing room, or a place of seclusion. Next, you need to break out of your comfort zone while still retaining that prepared or open mind. This is the only way you can meet and discover new people, talk to strangers, make new friends, and establish unexpected connections. When you travel, travel alone and let your curiosity guide you. Use this childlike curiosity as a compass, and with an open mind, you will be able to engage with new situations, places, and people. This will open you up to new things and encourage you to continue following your own instincts.

To simplify the whole process, imagine taking off your glasses (if you wear them); do not focus on a target, but just look as if you were NOT looking. Feel your gaze. Do not register what you see, but *feel* it. What is it that you feel? Look at what you are targeting from different angles, always with the goal of not focusing. Enjoy this holistic view of your subject; it will strike you if you are lucky and in the right mood. You will not find words in the process, but the images and feelings can be translated into something like a new order. Find something that needs to be added to the object or process you are working on. It could be something you have overlooked so far; at the right moment, you might discover it. This is the method by which, for instance, the trolley was conceived. Someone observed people in airports carrying their luggage while, nearby, a worker was using a board with wheels to transport something. Combine these two separate things, and it becomes one — the practical trolley. Very simple!

The more scientific model of serendipity describes the process as follows:

- To notice something, you need a trigger — a verbal, textual, or visual cue. This triggers the whole experience.
- There is often a delay between the trigger and perception, i.e., it may take some time to make the right connection between the trigger and perception.
- After the trigger and the (possibly delayed) perception comes the connection, the recognition of the relationship between the trigger and experience (or knowledge).
- After the connection comes action or actions aimed at making the best use of the connection between the trigger and experience (or knowledge).

- Then comes the positive effect of the whole serendipitous experience — a result, an answer to your question, or even a revelation.
- Then, you will recognize the unexpected or surprising element expressed in one or more of the random experiences.
- After a while (or immediately), you will come to understand and appreciate the random experience (serendipity) as such. Perhaps you will wonder about it; perhaps you will wonder why you and others did not come to the same miraculous realization earlier.

Now you understand how to make new and perhaps great hidden connections that could lead you to make a new type of device, technology, discovery, or invention. The serendipity process may be essential in our historic fight against the climate crisis. Radical thinking has changed the world more times than we can count. Do not be intimidated by the seemingly overwhelming abundance of problems. Simply trust that your new method of de-focusing will help you better understand the totality of the problems before you.

References

Colman, D.R. (2006). The three princes of Serendip: Notes on a mysterious phenomenon. *McGill Journal of Medicine*, **9**(2).

Encyclopaedia Britannica. (n.d.). Serendity. https://www.britannica.com/dictionary/serendipity

Gabler Wirtschaftslexikon. (n.d.) First principle thinking. https://wirtschaftslexikon.gabler.de/definition/first-principle-thinking-123085

Merriam-Webster Dictionary. (n.d.). Serendity. https://www.merriam-webster.com/dictionary/serendipity

PhilArchive! (n.d.) Serendipity as a strategic advantage. https://philarchive.org/archive/NGUSAA

Saplakoglu, Y. (2021). Sleep technique used by Salvador Dalí really works. Live Science, December 9. https://www.livescience.com/little-known-sleep-stage-may-be-creative-sweet-spot

Wikipedia. (n.d.). Horace Walpole. https://en.wikipedia.org/wiki/Horace_Walpole

Wikipedia. (n.d.). Hypnagogia. https://en.wikipedia.org/wiki/Hypnagogia

Wikipedia. (n.d.). Serendipity. https://en.wikipedia.org/wiki/Serendipity

Wikipedia. (n.d.). The Luckey Chance. https://en.wikipedia.org/wiki/The_Luckey_Chance

Wikipedia. (n.d.). The three princes of Serendip. https://en.wikipedia.org/wiki/The_Three_Princes_of_Serendip

Epilogue

If you want to change the future, you must change what you're doing in the present.

— Mark Twain (1835–1910), American writer

Climate change is arguably the most significant and serious challenge in human history. It has far-reaching and potentially catastrophic impacts on both the environment and global society. It already affects every corner of the planet and has led to a rise in global temperatures, as well as subsequent changes in weather patterns and sea levels worldwide.

Climate change has multifaceted, wide-ranging, and interconnected impacts, including hurricanes, floods, droughts, heat waves, rising sea levels, disruptions to ecosystems, loss of biodiversity, food and water scarcity, health issues, economic instability, and the consequent displacement of communities. Low-income communities in the Global South, indigenous peoples, coastal populations, and developing countries that lack the resources to adapt and recover effectively will be hit hardest.

The economic costs of climate change are immense, and expenses related to disaster response and recovery are skyrocketing. Changes in precipitation patterns and extreme weather events can disrupt agriculture and thus compromise food security. This leads to increased conflicts, displacements, and strained resources, which further intensifies societal disparities.

There is only a limited time for action. The urgency to act is paramount. Scale and speed have to be geared up significantly to meet the ambitious goals of the 2015 Paris Agreement. The longer we delay our actions to reduce emissions and adapt to changes, the more severe and irreversible the impacts will become.

To address climate change effectively, comprehensive global efforts, far-reaching policy changes, fundamental sustainable innovation, and international cooperation are required.

The Future of Climate Change

The future of climate change depends on a variety of factors. The most important are the future actions that governments, businesses, and individuals take to address the causes of climate change, mitigate its impacts, and adapt to the climate changes that have already occurred.

Fighting climate change requires a collective effort from individuals, communities, businesses, and governments. The most important actions you can take to make a positive impact on reducing greenhouse gas (GHG) emissions and mitigating climate change are:

* Reduce energy consumption and switch to using energy from renewables.
* Use less water, install water-saving appliances, and fix leaks.
* Reduce, reuse, and recycle. Minimize waste, reuse items, and buy second-hand.
* Change your common modes of transportation. Do not drive alone in your car. Instead, use public transportation, bike, or walk.
* Support sustainable policies; vote for climate-friendly politicians and policies.
* Eat and consume responsibly. Choose local organic food, reduce meat consumption significantly, and buy products made from sustainable or recycled materials.

* Support conservation activities and plant trees. Support reforestation and conservation.

* Reduce your use of plastics — no plastic bottles and use reusable bags.

* Support climate-friendly businesses by buying their products.

* Invest in green technologies, renewable energies, and sustainable agriculture.

* Engage in a community with sustainability goals. Participate in sustainable projects.

By adopting these practices and encouraging others to do the same, every one of us can make a significant difference in combating climate change and ensuring a sustainable future for all.

The impacts of climate change will depend largely on the extent and the pace at which global GHG emissions are reduced. Efforts needed to mitigate climate change include switching to renewable energy sources, such as solar and wind power, improving the energy efficiency of all types of technologies and machinery, rapidly implementing carbon pricing mechanisms, and adopting sustainable land use practices. The 2015 Paris Agreement provides the framework for international cooperation on climate change. The commitments made by countries under this agreement will play an important role in shaping the future trajectory of global emissions.

Technological advances, new and more efficient solar and wind technologies, carbon capture and storage (CCS), and sustainable agricultural practices will be critical for reducing emissions and mitigating climate change. Innovations in transportation, industry, and agriculture can help decouple economic growth from carbon emissions.

Climate change is already impacting all global and local ecosystems, communities, and economies. Adaptation measures of all kinds are essential in reducing the risks and impacts of climate change — measures such as building more resilient infrastructure, protecting coastal areas from sea-level rise, and developing climate-resilient agricultural practices.

Government policies and regulations will play a central role in addressing climate change. Countries and regions will need to adopt strong regulatory measures such as increased carbon pricing, more ambitious renewable energy targets, and harsh emissions reductions to meet the Net Zero 2050 target.

Climate change is a global problem that requires global collaboration. Only through collaboration at all levels, between the Global North and Global South and among countries of all forms of governance, will it be possible to build a resilient framework for robust further action.

Public awareness and engagement on climate change must grow steadily to become the driving force in influencing the actions of governments and businesses. Higher demand for sustainable products and services, as well as grassroots movements advocating for climate action, can drive change at all levels of society.

The future of climate change is also influenced by potential feedback loops within the climate system. Rapidly melting polar ice may further accelerate global warming by reducing the planet's reflectivity (albedo). Thawing permafrost could release additional GHGs, such as very potent methane, into the atmosphere. Melting of the Greenland ice cap could affect global ocean circulation with unpredictable consequences.

The future of climate change is uncertain, and many scenarios are possible. It depends on the choices that individuals, communities, and nations make in these critical years. The urgency to address climate change is evident when looking at the increasing frequency of disasters caused by these climatic shifts. The consequences of continued inaction or inadequate action include more severe weather events, faster sea-level rise, greater biodiversity loss, and local and global disruptions to food and water supplies. Thus, a variety of concerted efforts to reduce emissions and build resilience are critical if we have any hope of mitigating the worst impacts of climate change rapidly and massively.

The Bigger Picture

At its core, the climate crisis is an energy crisis, as the carbon footprint of human civilization is seen as the main cause of global warming by most scientists around the world. The ever-growing carbon dioxide level in the atmosphere is caused by burning fossil fuels that feed the global prime movers — internal combustion engines. We need to find new energy resources to feed our species' growing energy hunger. Our new energy resources must be sustainable, renewable, and "green."

The industrial sector plays a crucial role in both contributing to climate change through GHG emissions and putting new technologies into place that are based on sustainable, "green" technologies. To combat climate change more effectively, industries must adopt sustainable and environmentally friendly practices at a faster pace and on a much larger scale.

It seems that a growing part of the financial world has grasped the importance of this task. Investing in new technologies is the key to the survival of the global system that humans have created as "our world." In the near future, key energy technology could be hydrogen-based. There has already been a paradigm shift when it comes to energy transitions and adopting new future fuels for environmental protection.

There is a fast-growing trend in the development of the hydrogen market, and it is estimated that the production of hydrogen will grow 20-fold in the years from 2023 to 2030. Saudi Arabia (NEOM green hydrogen project), China (Ordos green hydrogen project), and Germany (FFI and TES green hydrogen project), as well as other large investments in green hydrogen projects in Finland, Australia, and other nations, clearly show the way of future developments. Moving from unsustainable fossil fuels to renewable energy sources is the only way to achieve a smart, reliable, and sustainable global energy supply — the backbone of our global economy.

The Kardashev Scale is a method of measuring how advanced a civilization's technological achievements are based on the amount of energy it can harness. Kardashev's paper, "Transmission of information by extraterrestrial civilizations," was initially focused on communication technology but has been expanded in a variety of ways by others.

A Type 1 civilization (or "planetary civilization") has the capacity to harness all the energy of its home planet. This means utilizing the energy that reaches the planet (i.e., solar) and all the energy it can produce (thermal, hydro, wind). In addition, this Type 1 civilization should be able to control things like earthquakes, the weather, and volcanoes, and it would be building ocean cities, according to renowned physicist Michio Kaku of New York University (NYU).

A Type 2 civilization can make use of the total energy potential of its star — in our case, the Sun. This stage — harnessing the Sun — would be a massive jump and might become a reality in a thousand years or so.

A Type 3 civilization is "a civilization in possession of energy on the scale of its own galaxy," as Kardashev saw it. This could be achieved in maybe 100,000 years. A Type 4 civilization could harness the energy of a whole universe, and we cannot even imagine or describe what Types 5–7 civilizations would be capable of. One thing is certain: Civilizations with high capabilities regarding their energy supply and use must be very old.

If you look at the Standard Model of Physics, it is very likely that there could be very old star systems with, accordingly, very old civilizations. These are called super-civilizations. Given the amount of exoplanets that humanity has been able to localize in recent decades, it seems increasingly likely that exocivilizations exist or, rather, have existed. The big problem with all these theoretical hypotheses is clear: Where are they?

The so-called Fermi paradox, named after Enrico Fermi (Italian physicist, 1901–1954), asked this question. The conflict is that scale and probability favor a huge likelihood of intelligent life in the universe, yet we have seen no evidence of it. "Someone from somewhere must have come calling by now." (Overbye, 2015)

If no one is out there, we could be the only ones to have survived the battle to become a higher civilization so far. The clear conclusion, if that is the case, would be that we are precious. That is why we have to do all we can to cope with climate change, find new energy resources (or rediscover old ones such as wind and solar), and strive for a global community of reasonable, peacekeeping, enlightened people.

Outlook

The world changes when an era comes to an end, and inevitably, all living people have to change accordingly. The special thing about change is that it represents a threshold to something new. The transition this time will be a disruption, as always, and an upgrade, as it always is.

A transition is also a change of mind. Crossing boundaries is scary, especially when whole systems, such as the old rules of capitalism and the even older use of fossil fuels, must be abandoned very quickly.

Our near future will be a transition to new levels of human intelligence, shared understanding, and global respect. It seems unclear who will lead us through the coming times; we will be largely dependent on ourselves. We will need to create something like a "shared intelligence," perhaps even with the help of new technologies or their working principles integrated into our own brains. All of this is but a glimmer on the horizon; stormy waters lie between then and now.

The sense of a common good is a kind of religious feeling. Without some kind of religious feeling, it will not be possible to keep our spirits up on the ardent road ahead. We are all sure that there is such a thing as a common good.

This book has told a brief history of mankind. There have been many great disasters, and there have been great breakthroughs, new inventions, and new ideas that established new ways of living together. This interplay of catastrophes and of new ways opening up runs through the entire history of mankind.

Several new inventions have been described in this book, along with some necessary steps to change societies and ourselves. We are on our way to a new territory, and we do not know where we will end up. The situation is similar to the journey Columbus and his crew made to find their way to India, which obviously turned out quite differently.

When there is a disruption, people get scared. We see disruption everywhere, and it is piling up. That is why so many people are feeling great fear in the world today. When you do not know where you are going to go and how you are going to go on, it can be paralyzing. The change you need to make now will be hard, and the road will be dirty. You will need all your willpower to get through it in good health and mental wellness. However, in the end, your brave and courageous journey in search of new realms of the human spirit will be worth the effort. Let us begin the journey together!

Figure 51.   Webb's First Deep Field, galaxy cluster SMACS 0723, NASA, ESA, CSA, and STScI.

Reference

Overbye, D. (2015). The flip side of optimism about life on other planets. *The New York Times*, August 3. https://www.nytimes.com/2015/08/04/science/space/the-flip-side-of-optimism-about-life-on-other-planets.html

Index

www.ingramcontent.com/pod-product-compliance
Lightning Source LLC
Chambersburg PA
CBHW050537190326

41458CB00007B/1815